Society Elsewhere

Why the Gravest Threat to Humanity Will Come From Within

T0154601

Society
Elsewhere

Why the Gravest Threat to Humanity Will Come From Within

Francis Sanzaro, Ph.D.

Winchester, UK
Washington, USA

First published by Zero Books, 2018
Zero Books is an imprint of John Hunt Publishing Ltd., Laurel House, Station Approach,
Alresford, Hants, SO24 9JH, UK
office1@jhpbooks.net
www.johnhuntpublishing.com
www.zero-books.net

For distributor details and how to order please visit the 'Ordering' section on our website.

Text copyright: Francis Sanzaro 2017

ISBN: 978 1 78535 470 0
978 1 78535 471 7 (ebook)
Library of Congress Control Number: 2017939489

A CIP catalogue record for this book is available from the British Library.

Design: Stuart Davies

Printed and bound by CPI Group (UK) Ltd, Croydon, CR0 4YY, UK

We operate a distinctive and ethical publishing philosophy in
all areas of our business, from our global network of authors to
production and worldwide distribution.

Contents

Perfection is achieved, not when there is nothing more to add, but when there is nothing left to take away.
Antoine de Saint-Exupery

It is a small part of life we really live.
—Anonymous Roman poet

This book is dedicated to Frankie IV.
I hope you prove the Roman poet wrong.

Introduction

Charismatic Code and the Frog

An urban legend has it that if you drop a frog in boiling water, it will jump out immediately. But if you put it in cold water and turn up the heat, it will be boiled alive. Despite the fact that this is not true, numerous experiments have proven its falsity, there is indeed a psychological truth to it—when so many little things change over a period of time, it's hard to notice. We are distracted. We are tired. We were busy. Then, it's too late.

If, in the last 24 hours, you have used your phone, watched something, read something, posted something or clicked on something, then you have been influenced by an algorithm. It sounds harmless, but it is not and we are proverbial frogs in, let us say, water just above room temperature...and the heat is rising.

It is also likely in the past week you have read an article written entirely by an algorithm or watched a movie whose success is largely the result of an algorithm, and, in both instances, you were not informed as to what role, exactly, an algorithm played. It kinda just happens, and it is happening more often in more places of our lives...and we go with it. We shrug. We're tired. And yet, it is a human problem, not a technological one.

Algorithms are, by far, the most powerful force in our lives today, both personal, political, medical, professional and even ecological, and we know nothing about them. In fact, we cannot know anything about them because they are protected by intellectual property rights and often are the foundation of some of the most successful companies on Earth, such as Facebook or Google. So, for Google to reveal its search results algorithm is tantamount to suicide, since once the code is made public, Google has no competitive edge. But people are beginning to demand to know more about these shadowy entities.

In October of 2016, German Chancellor Angela Merkel went so far as to say that some social media and news algorithms are politically dangerous. In the wake of a recent spate of terrorism and killings, Merkel warned, "These algorithms, when they are not transparent, can lead to a distortion of our perception, they narrow our breadth of information."[1] Merkel demanded such algorithms be made public. Of course, they won't be. In fact, Facebook's algorithms are now accused of being a major player in the 2016 US election cycle, not only for creating a "filter bubble" for its members, which can influence the extent of political engagement, but also for trending "fake news," which can effect just about everything.[2] None of this was intentional by Facebook, but still, politicians and users are beginning to demand to know what its algorithms are up to. Like Google, Facebook will reveal nothing. Yet, the larger point remains — how are culture and our lives changing because of algorithms? How are *we* changing?

But let us paint a broader picture. As of 2017, algorithms — which are formulas for telling computers what to do — are finding us our soul mates, writing news articles, judging fiction and film, controlling traffic, grading Hollywood scripts, tracking animals, composing music, writing for the AP news service, acting as mental therapists, telling us jokes, conducting job interviews, flying airplanes, farming without human intervention, driving cars, packing and shipping boxes, writing poetry that poets can't tell was done by a computer, running restaurants and, hot off the press, they are deciding what images are appearing on some major websites. In other words, algorithms are making inroads into the art and "creative" objects we consume — since the majority of images we look at on a daily basis are, let's face it, from websites. The list expands weekly. Non-human actors are modifying the texture, mood and fabric of our practical and creative cultures with an intensity unparalleled in history.

If you have ever read anything about the coming "automation revolution" —when machines take our jobs and rule the world, and we sit back helpless (which is unlikely)—you are reading indirectly about the power of algorithms.

So, what is the threat here? *Uselessness.* We are rendering ourselves useless, in terms of not feeling necessary, feeling out of control and/or sensing that one is without utility and without an expectation of fulfilling an intended purpose. Uselessness is a shape shifter, roaming around our consciousness, wreaking havoc, morphing into boredom, lack of engagement, aggression, instability and decreased awareness of one's surroundings. When the latter are unleashed, politics, social life, addiction rates and our romantic lives are upended. This threat is real, costly, irrational, uncontrollable, and comes with the potential to release profound amounts of psychological disturbance into our psychic (and actual) economies. Future politicians, therefore, will be tasked with much more difficult assignments than border disputes or outbreaks of disease, but rather the management of the emotional lives of hundreds of millions of people who lack the basic sentiments we have always taken for granted—being in power, being in control, having a sense of destiny or purpose. Israeli historian and thinker Yuval Harari rightly observes that the "economics and politics of the coming decades will be what to do with all these useless people."[3]

Boredom, as one of the fastest-growing "afflictions" affecting modern societies, has been directly tied to how technology is changing our habits of perception and attention. As you have probably guessed, *feeling* useless is a main trait of *being* bored. Uselessness, therefore, is an existential crisis. This is not a prophecy, but if you follow me closely enough in this book, I believe you will find it to be quite a logical outcome, and rather terrifying.

Science and Shit

I can hear you saying, *Really, uselessness is the future's biggest threat? Come on.* But first, the obvious question—why aren't hunger, demagogues or ecological collapse the biggest threats to our future? If something threatens our future in a tangible and existential manner, surely it is hunger, or maybe nuclear war. Right? Well, political and technological progress up until this point has not given us a reason to believe this will be the case. With every historical rise in the global population, we have met the threat of food shortages with remarkable innovation, and countless predictions of global starvation and calamity have just not panned out. As researchers Erik Brynjolfsson and Andrew McAfee have recently argued, technological innovation (beginning with the Industrial Revolution) is responsible for the most profound instance, and rate, of global transformation this Earth has ever witnessed.[4] Not only did the Industrial Revolution lead to exponential population increases, but, in the midst of seemingly insurmountable challenges it was able to, to quote Matt Damon's character in *The Martian*, "science the shit out of it." In other words, technological innovation is proving to be better than we thought at solving its own problems. This is not to trivialize the very real threat nuclear wars and exponential population growth will put onto our societies, only that history has proven we are up to the challenge.[5]

"I Am Everywhere"

Again, why uselessness? According to many industry insiders and researchers, the coming decades are being heralded as the "age of the algorithm," and a major 2013 study out of Oxford University states that nearly 47 percent of US jobs—not just blue-collar jobs—will be replaced by smart computers running sophisticated algorithms within the next two decades.[6] Around 40 percent of jobs in New Zealand are said to become irrelevant in the near future, 35 percent in the UK and about 50 percent of

Indian IT jobs. Search "jobs" and "automation" and you will see that lawyers, financial analysts, educators, doctors, musicians and writers are already being replaced by algorithms. The same Oxford researchers revisited the topic in 2016, this time revising their numbers for the worse. Automation will affect countries and cities in varying degrees, but it is unanimous that a momentous shift is coming, like a tsunami born in the middle of the Pacific—it is only a matter of time before it washes onto Ventura Beach.

Everywhere there seems to be a silent crisis being called out from the insides of nearly all industries, despite the fact that automation, in all its historical manifestations—including the automated machinery of the Industrial Revolution—was supposed to be liberating. Noted visionaries, such as Oscar Wilde, or Karl Marx, said as much. Automation was supposed to free up our precious human time and let machines do the dirty work that did not require higher cognitive skill, while we reaped the benefits of increased leisure. However, as we all know—in highly sophisticated societies where automation is firmly entrenched, we are en masse working longer hours for either decreasing or stagnate wages. In fact, the average American worker only cashes in on half of their vacation time and 15 percent do not take any holidays at all.[7] The more techno savvy we become with devices intended to save us time actually translates into longer working hours. Furthermore, we are in increasingly what anthropologist David Graeber has referred to as "bullshit jobs"—managerial, clerical, sales and data services workers for whom the daily tasks are meaningless.[8] It goes without saying that a person with a "meaningless" job feels useless. Soon, even the bullshit jobs could be gone.

Do Not Resuscitate

This book *is not* a journalistic enterprise about the history of how human habits have changed with each new device, and nor is it a focused study on how automation and its close kin—artificial intelligence, robotic machinery, etc.—is altering the nature of our workplace; this is not a history of technology. There are plenty of excellent books on those topics; see the references at the end of this book for some guidance. This book *is not* a "let the facts speak for themselves." Finally, this book is *not* a glorification and a gee-wiz embrace of the computing revolution, digitalization and/or the hundreds of ways software is finding itself embezzled into the nooks and crannies of our lives. Rather, this book is about humanity, you and I, the things we love, our children and our happiness.

For anyone who has had to deal with elderly parents or grandparents, one of the primary tasks to do while they *have the capacity to think* is plan for a time when *they no longer have the capacity to think*. It is a strange thing to do, requiring legal documents, contingencies, official forms and notaries, but it is a reality and it is done every day. What is not done, by virtue of being much messier, is planning for a type of cultural incapacity when it is much more likely than not that in the future said culture will not resemble itself, will be unable to make good decisions about itself, and so is in need of some type of baseline "intelligence" to judge itself against. In the case of elderly parents, the baseline intelligence is the sons or daughters, since the duty of responsible thinking has passed from the parent to the kin, and this transference functions like a living archive of the individual's desires and wants, such that when the time comes, one can confidently say "they wouldn't like to live like this," and then the hard decision can be made: Do Not Resuscitate. This book is not just trying to establish a baseline intelligence in an era when we are outsourcing ours, but rather to paint a portrait of cultural and natural health in the process.

2015 Francis Sanzaro, Ph.D.

Part I

Humanity

Chapter 1

More Nude Scenes

Man, in the large, is a mass urged on by a force.
—Tesla

Algorithms are amazing, vicious, poetic, ruthless, stupid and genius. As the DNA of modern computers, they are elegantly simple, exceedingly complex and central to the apps on your phone, artificial intelligence and all the software or devices offering suggestions, filtering, sending notifications and/ or making complex decisions for you. In short, an algorithm has directed your life today. It is only natural, therefore, that a seminal book on software studies, which goes by the same title, has "algorithm" as its first entry.[1] These subtle lines of code dancing on the programmer's page are the past, present and future of computation, and they are the one thing computer scientists agree upon—"Without the algorithm...there would be no computing."[2] Existing in mathematical models and in abstraction, either in our brains or on paper, an algorithm comes alive when it is embodied in computer hardware; they have to be materialized to have an effect in this world.[3] Just as a Hollywood script needs actors for it to become a play, in order for an algorithm to flex its muscle it needs data (inputs) and, as we are all now aware, data can be anything.

According to Thomas Cormen, author of *Algorithms Unlocked*, "Computer algorithms solve computational problems. We want two things from a computer algorithm: given an input to a problem, it should always produce a correct solution to the problem, and it should use computational resources efficiently while doing so."[4] In other words, algorithms are a formula or procedure for telling computers what to do, and need to be consistent and efficient in their use of data when providing a solution. Computers are algorithm machines.[5] Consider the rest of this chapter as a type of algorithm—except the "input" is the algorithm itself, being fed into culture, and the "output" being its effect on our cultural health.

"More Nude Scenes"

Algorithms help Google Maps get you home when there is an accident ahead and you need to get off the highway. Do you go east off the highway? Do you go west? Now that you have gone east off the highway, what street to take next? The first street on the right has a lot of lights down the road, but the third left is a meandering single lane prone to build-up during rush hour. Is it rush hour? What is the weather? What happened last week? What is the trend for this route—has congestion been increasing or decreasing over the past month, week or day? What is the data being tracked from other drivers' phones saying? All this information (inputs) is fed into algorithms to get you home the quickest (an output in the form of a route). It is a simple process, but human minds cannot process a fraction of the data as quickly as your GPS, which is why computers started beating human players in chess (in the late 1980s) the moment they had more speed and processing power. Many experts thought technological progress would claim its advancements on the back of processing speed, Ray Kurzweil, now Director of Engineering at Google, being a famous proponent of linking processing speed and capability, but algorithms may be the true revolution. In one report, drafted by the US President's Council of Advisors on Science and Technology and titled, "Report to the President and Congress: Designing a Digital Future: Federally Funded R&D in Networking and IT," the following observation is made— "Even more remarkable—and even less widely understood—is that in many areas, performance gains due to improvements in algorithms have vastly exceeded even the dramatic performance gains due to increased processor speed."[6]

Output of picture gallery one: our decision-making capabilities regarding geographical orientation and awareness are being eroded.

Algorithms are allowing political parties to cater their marketing material specifically to your neighbors.[7] For instance, did your neighbor lose their wife recently? Well, big data can predict what type of political narrative will most likely appeal to your grieving neighbor, perhaps a narrative of empathy and compassion with tidbits of injustice. Social profiles of us are already available for purchase on the Web and political parties can purchase those profiles, feed that data into their systems and spit out brochures, robo calls and emails "perfectly" tailored to your existence.

Output of picture gallery two: the political intelligence needed to understand issues in communities has given way to data tracking and automated analysis, thereby rendering some parts of the "boots on the ground" political process useless. The only logical result of this trend is voter distrust, cynicism and anger, which will change the face of politics in each sector where this is implemented.

Algorithms are not only making inroads into politics on a marketing level, but on a policy level as well. Valentin Kassarnig has managed to develop a program able to write political speeches that are, apparently, difficult to determine if an underpaid staffer penned it or a computer wrote it. For instance, it produced: "Mr. Speaker, for years, honest but unfortunate consumers have had the ability to plead their case to come under bankruptcy."[8] How did Kassarnig do it? By using an algorithm to scan and analyze 4,000 political speeches, find patterns, and then produce results implementing those patterns. Once the algorithm knows what is effective and going on in speeches already written by human minds—word length, affective tone, frequency of verbs used, etc.—it can clone that pattern and output an analogous speech, and as you can see it does a fairly good job. Users of Gmail now have the option, called Smart Reply, of semi-automated email responses, wherein a program analyzes an incoming email and then offers a few responses, replete with the right tone and all.[9]

Output of picture gallery three: the skill set to construct public intellectual products such as political speech will render an entire class of political intellectuals useless. While this is a small example, when applied elsewhere the ability to construct meaningful sentences for public consumption will become a useless skill, along with the thought process needed to create such speech.

When algorithms and pattern/association matching processes are brought together you get apps making decisions for us, apps based on algorithms that, when combined with other soft and hardware, gives rise to something feeling like artificial intelligence. I am referring to the culture of the playlist "Recommendation" or "Because you watched..." A more complex and recent phenomenon regarding content creation, the entertainment industry already uses algorithms to predict the box office success of a film, an algorithm using characters, plot and actions as input "data points." Epagogix, the developer of this "smart" script reader, describes their product in the following manner:

Advanced Artificial Intelligence in combination with proprietary expert process enables Epagogix to provide studios, independent producers and investors with early analysis and forecasts of the Box Office potential of a script. Clients then make evidenced decisions about whether or not to spend their scarce capital, adjust budgets, or to increase the Box Office value of the property.[10]

In other words, their algorithms can tell a producer that their film starring such and such an actor will only pull in 20 million as opposed to the 35 million the studio thought it would...before the filming even begins. With the help of Epagogix the production team can decide to get a more attractive, lead female character (a bigger net gross), add more nude scenes, or keep things as they are and pay the actress less, since they are confident the film will underperform and their ROI outlook is now revised.

Output of picture gallery four: our ability to cultivate a personal aesthetic is becoming unnecessary. Moreover, the creative individuals who read, edit and review such materials will soon become useless, as the act of reviewing and editing will become automated. All you have to do is combine picture gallery three and four.

Algorithms are beginning to determine what books get published from public domain. January 1st of each year is called "public domain day" for many publishers, as this is the day the copyright sticker on books written decades ago is torn off. One researcher has found out a way to mine the views, clicks and article lengths of Wiki pages, in combination with data from a database of books already in the public domain, and, when put together and processed accordingly, a ranking can be generated.[11] Publishers can then use this ranking to decide on which books to publish.

Output of picture gallery five: the intellectual industry of publishing that looks to great works of art and cultural trends to publish books for public consumption is becoming obsolete, with effects that will trickle down not only to a decreased appreciation in books and cultural teachers, but also in education programs that hone this skill.

We now have a coat that hugs its wearer—from behind—when it detects said individual needs a hug and is openly termed a substitute girlfriend; to be fair, it is only a prototype at this point. The wearer connects to headphones and along with a hug, there are prerecorded emotional voice bits such as, "I'm sorry, were you waiting?" "Watch your back!" "Guess who?" "Blind side!" The fact that the coat hugs from behind, so as to catch the wearer by surprise, like a typical girlfriend would do (?), can, with little stretch of the mind, be best termed an existential cough drop—akin to professional cuddlers or services that send girlfriend'esque text messages to men who feel lonely—a bit of false humanity to stave off modern *anonyme*. And, given that the voice is from Yu Shimotsuki, "known for work in adult-themed video games," there is a seductive porn edge to it all.

> *Output of picture gallery six*: products are slowly chipping away at social initiative or the motivations thereof, situating themselves between our biological signals of, for example, loneliness and their solution, effectively, nay potentially, rendering ourselves useless in the process of knowing our emotional states and modifying that emotional state.

Lest we get the impression that algorithms are trivial entities, they are being deployed by over a dozen young startups to fight cancer and ultimately could save thousands of lives in the years to come. IBM, BERG Health and CureMetrix, to name a few, are using AI algorithms to interpret body scans for cancerous growths, to detect minute changes in temperature from a woman's breast tissue so as to raise an early warning sign of breast cancer, and to help companies, such as Globavir Biosciences, reduce the time it takes to develop new drugs.

Output of picture gallery seven: the skills of professional medical researchers are being enhanced, not replaced, by AI algorithms; however, the utility of the researcher will diminish the more sophisticated our software becomes.

Related to the hugging coat, there is the T.Jacket, about $549 USD at the time of this writing, which is a jacket worn by kids allowing parents, via a network of air pockets and sensors, and, of course, an app, to hug their kids remotely. Johnny gets upset and smacks Cassandra at school. Johnny's pulse and body waves indicate he is distressed. The jacket, set by the parents to activate at certain thresholds, inflates and hacks the body-love system, that is, intervenes between biological response and stimulus, and, ideally, calms the child. The invention is backed by research about touch therapy for certain individuals, as well as by certain forms of mental conditions, where pressure on the body significantly helps the individual feel secure. Also, software is beginning to decipher the cries of an infant. Is your 2-month-old baby hungry or distressed? Who the hell knows most of the time!? They are two months old. No one knows. Yet, the sound waves and types of cries are treated as inputs into software, and the output is an interpretation of your child's fully inarticulate yelps and gestures.

As you can see, we are well beyond functional algorithms: they live between us and our instincts, between parent and child. How far of a stretch is it to imagine that, in the future, when the software detects a hungry infant, a machine automatically makes a bottle, and, for the adventurous types, a robo-maid delivers the bottle to the child. Do we applaud this move, since now mothers and fathers, so pressed for time, can focus on something other than guessing why their infant is crying and holding them or swaddling them or rocking them to no avail? Or is this a Trojan horse, with the inevitable result of eroding basic human bonds at one of the infant's most vulnerable times in their lives?

Output of picture gallery eight: software is being situated at intimate levels at stages of human bonding and can render the parent's instinct useless, or at least degrade their confidence in listening to their children. While this is a small example, the applications for this technology in all parts of behavioral and speech recognition are tremendous.

There are belts to help with overeating by analyzing our biometrics and generating a real-time diet, and Microsoft has developed a bra to help women make better food decisions as they near the fridge seeking a treat.[12] Another company, this time Japanese lingerie brand Ravijour, has created a bra set to unhook only when its sensors indicate that, indeed, the wearer is in love. Fundawear, a product developed by Durex, is allowing users to play digitally with their lover's body — yes, in those parts — by an app. INTIMACY, a line of avant-garde fashion clothing coming out of Studio Roosegaarde, can make its fabrics transparent by reading the data on the body.[13] So, presumably, based on the signals it is receiving from the wearer's body, your clothes turn transparent. I can only imagine the hacking potential here on Friday night at the bar, surely to become another form of digital sexual assault.

On some software, it is possible for women to log their menstrual cycles, to help them better define the times when they are most fertile. In other words, an app can now tell you when to have sex.[14] Another company has patented an electronic orgasm delivery devise that is apparently only for medical purposes, such as for women with "orgasmic dysfunction."[15] Able to deliver an orgasm with the push of a button, we should believe this machine will only be used by the dysfunctional with as much credence as we think the Internet is only used as a tool for enlightenment.

Output of picture gallery nine: while seemingly trivial, the presumptuousness of software to hack our ability to read our bodies at the level of "love" is aggressive and somewhat humorous, like a gag gift from the mall, but the use of diet software will cause a decrease in our confidence to read and understand ourselves, make informed decisions about how we consume food and for what reasons. Akin to the jacket that automatically hugs its wearer, the orgasm delivery device is further grease on the wheels of social alienation and personalized pleasure delivery.

Even the highly nuanced procedures of legal analysis and documentation, it appears, is under the threat of automation. While a trove of headlines lamenting lawyers will soon be replaced by software able to sift through, find and analyze the relevant cases for determining a defense, white paper or procedural strategy, movements have already been made in this direction.[16] The discipline of psychotherapist evaluation, often labor intensive, could give way to programs listening to patient-therapist conversations, transcribing the conversation to text, analyzing the text for "counselor empathy" and then providing a rating. Apparently, early efforts of software to detect counselor empathy against traditional human interpretations have proved close.[17] Naturally, it is only a matter of time before the software is offering suggestions to the therapist to make them more empathic. Moreover, once the metrics from the social situation or whatever are made available, real-time, to the software, it will further be able to predict the patient's mood, then send tips or directives to the therapist in real time for implementation in a session. Say a patient is a good liar and, though seemingly aggressive, is actually falling in love with the therapist (a common problem), then cautionary measures can be taken immediately; perhaps the algorithm made this determination based on body heat, body language, doublespeak and eye contact, wherein it might have taken the therapist a few sessions to realize. In a boost to Google's self-driving car ambitions—and the desire to humanize software—in 2016 the US National Highway Traffic Safety Administration said that it considers the "self-driving system" to be a "driver."[18] Analogously, software will soon be considered a "therapist."

Output of picture gallery ten: algorithms are making complex social and empathic skills, and their training, increasingly useless—deep listening, legal rationality—since automated lawyers and therapists are cheap, do not complain, and can be given to millions of people simultaneously, like any digital product. Along with this development, varying types of intelligence will become dulled and based less on human-to-human interaction. Algorithms are becoming personas and given that Google's software is considered a driver, why shouldn't software be given the title of "therapist" or "lawyer"?

Nursing homes are dabbling with robot caregivers and law enforcement is dabbling with machine learning to predict, for example, which domestic abusers will likely be re-arrested.[19] Though seemingly in its infancy, algorithms are helping some law enforcement districts predict, for example, if an individual is highly likely to be rearrested on the same charge within the near term. If, based on a set of over thirty-five characteristics, such as age, gender, prior history of charges and so on, the algorithm determines a strong percentage of recidivism, then it seems "reasonable," after a human mind has analyzed the results, not to offer that individual parole.[20] Likely, this type of predictive policing will be scaffolded into the system slowly and assuredly, with strong oversight and quality control, at least at first. When detractors argue such an incorporation will only reinforce stereotypes, reduce freedom and put us all under the yoke of machines, the counterargument will be that "such and such" software helped catch "such and such" number of individuals before they were able to abuse children. And, the detractors will have a point.

> *Output of picture gallery eleven*: algorithms will find themselves in greater positions of power in all facets of the legal system. Akin to the intelligence of detectives, the role of officers and their mental abilities will slowly decline until they themselves become monitors of the algorithms. Human intellectual capital within the justice system is in danger of becoming useless; though I imagine this sector of culture will be the last holdout for handing over real power to algorithms.

If This, Then That

The eleven picture galleries are just a snippet of what algorithms are doing today, yet many of them are part of a structural change in how we participate in our surroundings. That structural change can be summed up as a personalized world and Eric Schmidt, Executive Chairman of Google, described it thus: in the future, observes Schmidt, "There will be so many IP addresses… so many devices, sensors, things that you are wearing, things that you are interacting with that you won't even sense it."[21] Schmidt continues: "It will be part of your presence all the time. Imagine you walk into a room and the room is dynamic. And with your permission and all of that, you are interacting with the things going on in the room."[22] Just as the trees in a forest bend towards the sun, the near future will make objects "bend" toward us via software.

IFTTT, short for "If this, then that," is a relatively new app allowing users to bundle their gadgets and app services by working on the basic premise of algorithms — single command directives. For instance, if I do this thing, then I can tell the app to do another thing. IFTTT has honed the trickle-down effect of layered media and it will be a crucial inspiration to future biometrical automation which, for instance, could dictate that whenever I enter a place, even if I have not been there before, the lights dim to my liking, food will be ordered without my intervention and a drink will appear just the way I like it. The future is personalized, all aspects, just the way "targeted" advertisement attempts to be.

All public spaces in the future will deploy this technology, most likely in increments. Already in South Korea, Dunkin Donuts began what *The Atlantic* termed the future of advertising—spraying donut-smelling fragrance into public spaces, such as buses and subway cars, under the assumption that bypassing your intelligence and going right to the pleasure centers of your body is the most efficient way "in." Now, what if chips in the subway could detect that *you* are sitting there—easy to do because of the personal profile your smart phone can broadcast and your unique IP address—and then you would get personalized marketing smells catered to you. If this sounds out there, it is not. They could even create the smells of your teenage years, gathered from your credit card purchases and the types of lotions and creams you bought when you were fifteen. Maybe they play on the speaker an ad in which the voice sounds eerily like your grandmother—all phone calls are recorded—because it is a delivered via a voice-matching API that is actually using your grandmother's voice, or a close rendering, which was taken from phone calls you had with her years ago.

∞

Civilizations have undergone similar automated transformations before, notably the Industrial Revolution, but this time our coming personalized worlds have no precedent, since then only factory labor was affected, along with the economics thereof. Undeniable is the increasing use of smart algorithms to guide humanity and enhance humanity, and this takes a variety of forms and strategies—getting in between ourselves and our bodies, between one another, aiding in "thinking" or research tasks, or allowing us not to worry about some tasks altogether. If our brains can be imagined as a box of tools that we get out to use when we go about in the world, we are more frequently leaving those tools in the box. The ramifications of that are profound and it should keep us up at night.

In the end it is a balance, right? Some applications of algorithms are good, some trivial and some plain old creepy, if not downright intrusive. But this is just the beginning of how to think about their role in our world and, as it turns out, the promise of technology to guide the progressive hand of "culture" to ever greater heights has its own history, replete with fantastical incantations, dreamlike predictions and images of humanity lollygagging under trees while machines do our work for us. As a population allowing software into our bedrooms and underwear, we have to date failed to discern why we are letting this happen. It is not circumstance, nor predetermined by technology, since algorithms have thousands of applications other than doing things for us, just as a quill can be used to pen a constitution or gouge out an eye. But we have chosen this path for their use and so the question is—*what does this tell us about ourselves?* In the same way you realize as an adult why you kept dating abusive women in your twenties, it is a question of deep human peculiarities, of being promised to, scared, overjoyed or in denial.

Chapter 2

History of a Fantasy

Throughout history, poets, artists and philosophers have painted a picture of the human mind as a balloon submerged in water, held down by strong hands. The hands are duty, responsibility, lack of time and the fact that we are no longer nomadic hunter-gatherers, which means we need to trade our labor for life's basic necessities. But, if given the opportunity and the hands let go, the balloon would instantaneously shoot up from the depths, rise out of the water like a newborn and career toward the heavens to capture its ethereal bliss. However, during most of our free time we get drunk, swipe and tap, spend recklessly or get the hell out of the places we currently live in frantic rush, all in order to "relax."

∞

Basic Stone Age tools, such as the ape putting the stick down the ant hole, are forms of early automation, since the stick "did the work" for the animal. Hunting traps are a form of automated intelligence as well. Automation, tools and the ultimate aim of humanity—it is a strange pairing. Del Harder, a former VP of the Ford Motor Company, brought us the word "automation" in 1947, a term overshadowed by the revolution it was trying to describe. In stark terms, he defined it as the "automatic handling of parts between progressive production processes."[1] This definition is only of interest in contrast with what automation has become today, since no one would consider your email's junk folder a form of automation and a direct descendent of the seventeenth century, but in truth it is. Del Harder did not invent automation, but put the word into our lexicon.

As with most things, the first hints of industrial automation elicited praise in addition to fear, utopian dreams along with dystopian realities. A Janus-faced reality. After the Industrial Revolution of the eighteenth and nineteenth centuries, we can credit the Luddites for being the first to consciously eschew machinic production, because it was believed automatic machinery led to a poorer quality of living, unemployment, inequality, hunger and loss of craft. As primarily a labor movement, the Luddites were railing specifically against the introduction of automation into textile manufacturing; having spent years and sometimes decades mastering a craft, they were not going to allow a machine to replace their hands and years of training. In an effort stop the march of the machines, they proceeded to smash and set ablaze the machinery in select factories, beginning in 1811. The British government in turn made it punishable by death to break machines. Quickly, under the threat of death, the movement died. It is common to claim this group is the direct antecedent of technophobes. However, the Luddites were not technophobes, they simply did not want machines to replace human skillsets. They were only "technophobes" by default, as it were, since it could have been another market force that achieved the same "goal" as the machines. It just so happened that automated technology rendered their skills and intelligence unnecessary, and so they railed against the cause. Still, the general contours of the Luddites' anxieties extend into the twenty-first century anytime we talk about how machines—computers, devices, apps—get between ourselves and things traditionally considered the province of humanity.

In contrast to the perceived threats of automation—found not just in the Luddites but in Western philosophy, the arts and culture in general—was a perceived means to achieve the destiny of humankind by way of automation: the life of leisure, contemplation and joy. In many ways, each side presented a different vision of humanity and, as a result, the visual arts became a battleground, as it so often does, for competing visions — glorious urbanism and mystical smokestacks promoting the lure of progress and cafe-lined boulevards and flaneurs walking turtles on a leash to view the latest "mall" *versus* the noble peasant sowing seeds against a setting sun, haystacks, scenes of muscled farmers or boots painted with crusted mud, still slightly damp from the day's work. Urban leisure and the fruits of advanced civilization *versus* old-world contact with forests and sunshine. But of course, nothing is ever this black and white.

Writing in 1891, some fifty years after the period of rapid industrialization, Oscar Wilde, an influential Irish poet, essayist and novelist, presented a famous case in favor of automation:

All unintellectual labour, all monotonous, dull labour, all labour that deals with dreadful things, and involves unpleasant conditions, must be done by machinery. Machinery must work for us in coal mines, and do all sanitary services, and be the stoker of steamers, and clean the streets, and run messages on wet days, and do anything that is tedious or distressing. At present machinery competes against man. Under proper conditions machinery will serve man. There is no doubt at all that this is the future of machinery, and just as trees grow while the country gentleman is asleep, so while Humanity will be amusing itself, or enjoying cultivated leisure – which, and not labour, is the aim of man – or making beautiful things, or reading beautiful things, or simply contemplating the world with admiration and delight, machinery will be doing all the necessary and unpleasant work.[2]

Wilde's words are seductive, beautiful, appealing to our better, if not aristocratic, senses. The paragraph, whilst speaking of the dirty, oily processes of nineteenth-century industrialization, is about trees, fresh air, sleeping and contemplation. In his thoughts, Wilde had nothing but intimations of what automation could be in virtue of what it was then, and we cannot hold it against him that he did not live for another fifty years and see what really became of automation. For instance, note how at the beginning of Wilde's quote, he specifies that "unintellectual labour" should be handed to machines. While most likely Wilde could not imagine how machines *could* do intellectual work during his time, it is probably fair to assume that were he alive today, he would decry machines doing such tasks as composing music, writing poetry, judging fiction or writing the news, some of the things humans were supposed to be doing while machines were doing the other stuff. Wilde would turn over in his grave to learn that AP's Automated Insights, without human intervention or editorial guidance, penned the following story that begins:

> SANTA CLARA, CALIF. – Intel Corp. on Tuesday reported third-quarter revenue and profit that were down slightly as the giant chip-maker continued to confront declining demand for personal computers.
>
> But the company managed to beat Wall Street estimates for earnings and sales, which caused a brief jump in the company's share price in extended trading Tuesday. That gain was later erased and the stock slipped about 3 percent.[3]

It could also be the case that what we thought all along *was* intellectual work *wasn't*, but that is another question for another time. For the rest of us—will automation and AI help us in the task of "making beautiful things, or reading beautiful things, or simply contemplating the world with admiration and delight"? Wilde's analogy of trees growing while the countryman is asleep is quite eloquent and could be interpreted to mean that machines help humanity "grow" (the tree) while we can get some shut-eye under an English oak. Or, the analogy can mean we are growing like the tree while machines put life to "sleep"; for example, life can be put on autopilot. The sentiment is seductive—if only we could automate enough tasks, we could all live in a constant state of contemplation and wonder; that, if finally it can stretch its legs, the old philosophical dream to go about the world in a state of wonder (Plato's contemplation of the *good* or Aristotle's *theoria*) and awe would be achieved, and we could live as humanity ought to live—sleeping, making, reading, contemplating. This naturally begs the question of human purpose—Is leisure the aim of life? But let us not be seduced ourselves into the path of great questions, as they have a habit of not being solved and for missing the most important parts, mainly owing to human insecurities built into the questions themselves.

Famed electrical engineer and futurist Nikola Tesla, to which the current company using his last name honors, echoes the liberatory potential of automation in 1937, about forty-six years after Wilde:

Today the robot is an accepted fact, but the principle has not been pushed far enough. In the twenty-first century the robot will take the place which slave labor occupied in ancient civilization. There is no reason at all why most of this should not come to pass in less than a century, freeing mankind to pursue its higher aspirations.[4]

In 1930, seven years prior to Tesla, the British economist John Maynard Keynes, whose thoughts founded a branch of economic theory—Keynesian economics—made the following claim, "Thus for the first time since his creation man will be faced with his real, his permanent problem how to use his freedom from pressing economic cares, how to occupy the leisure, which science and compound interest will have won for him, to live wisely and agreeably and well."[5] In 2005, Ray Kurzweil stepped into the folds of historical dreams when, echoing Wilde and Keynes and so many others, he made the following prediction about how technology will affect our future: "The role of work will be to create knowledge of all kinds, from music and art to math and science. The role of play will also be to create knowledge. In the future, there won't be a clear distinction between work and play."[6]

Marx on Automation

Karl Marx, whose communal-based thought would place the workers' humanity at the center of new societies, bemoaned the "lifeless mechanism"—the machine—not necessarily for its threat to workers, though that was part of it, but for the existential pivot that occurs when you place something between the human body and the outside world: we begin to live for the machine. Marx observes: "In the factory we have a lifeless mechanism independent of the workman, who becomes its mere living appendage."[7] Marx believes that factory work (not to be confused with automated factories) "confiscates every atom of freedom, both in bodily and intellectual activity." But, according to Marx, automated factories may provide relief to some forms of alienation via time—automation can make possible "a large quantity of disposable time."[8] Automation could potentially lead to a new "realm of freedom"; for example, socialized man, assuming that additional time would be spent on education and the arts.[9]

On the other hand, as automation generates new forms of leisure, the worker becomes increasingly irrelevant and Marx's predictions vacillate between hope in automation, ala Wilde, and pessimism, knowing that it is an essential risk to remove labor from an individual's essential constitution, especially when the entire thrust of automation is to maximize profits. But, as we know, there are a few flaws in this train of thought as humanity's default position vis-à-vis "disposable time" is not to fill it up with soul-bolstering education and the arts, but to fill it up with distraction and the products of late capitalism.

Secondly, automation does not simply remove a task from life, but rather changes the nature of the task. As contemporary life has proven with large amounts of our tasks now automated, we find ourselves working longer hours so as to manage all of the automated tasks, which are increasing by the day.

If Marx was skeptical then he would be even more so today, with increasing numbers of people tied to automated calendars, aggressive notifications, automated driving routes you do not want to deviate from—even when the map leads you into the heart of the Nevada or a guardrail—monitored time-on-tasks and even app-based micro-tracking of an employee's every move and keystroke, the latter the type of millisecond monitoring that got Third World sweatshops the attention of human rights groups.

In our era, however, automated tracking and a good number of "personnel management" solutions are used to increase employee efficiency, which is the precise opposite of how early witnesses to automation envisaged its effect—it was thought we would be working less and have more free time, but today we are working longer hours and have to cram more into the hours we do work. If you have Internet restrictions at work, then there is, quite literally, automated software preventing you from "daydreaming," loosely conceived, despite the fact that with our modern tools we are some of the most productive employees to have graced the planet. It is only a few logical steps until we allow our personal and emotional lives to be monitored as well, since we will have experienced monitorization at work or elsewhere and it will just creep in, pseudo naturally. One thinks here of Microsoft's elusive Personal Agent project, which according to Bill Gates "will remember everything and help you go back and find things and help you pick what things to pay attention to."[10]

While history is no stranger to the notion of controlling employees, what is perhaps unique in this case is the Trojan horse phenomenon—under the guise of freedom, ability and efficiency, we gain restriction and the inability to be efficient, which is a real loss; just look at how much children's happiness comes from being inefficient, and we can learn something there. Lastly, and strangely, we will also become more well-behaved— since we are always being recorded and under the threat of very real consequences for behavior that historically wafted into the clouds—and we will take less risks, though outbursts of criminality and irrationality will become more common, from pent-up aggression. In the end, irrational splashes of cruelty is a very likely result of ceaseless monitoring in our professional and personal lives.

The Automation Ethos

In 1930, Keynes predicted the coming existence of a 15-hour workweek owing to "technological unemployment." While future unemployment was a problem for Keynes, figuring out what to do with our newfound free time was the bigger issue. He did not, like Tesla or Wilde, have enough faith in humanity that purposeful leisure is simply what you get when you take labor away—the balloon submerged in the water analogy—as if the default mode of humanity is comfortable bliss or time well spent. According to Keynes, for millennia our instincts have been hardwired to solve the "economic problem"—hunger, subsistence, etc.—and, moreover, since technology is poised to solve this problem (in the 1930s), humanity's main issue is how to stay purposeful, connected and relevant in a world where the traditional anchors of fulfillment—for example, work— are gone. Keynes worried about a mass "nervous breakdown" that could inevitably result, but struck a positive note when he observed, "But it will be those peoples, who can keep alive, and cultivate into a fuller perfection, the art of life itself and do not sell themselves for the means of life, who will be able to enjoy the abundance when it comes."[11]

If Keynes stressed about the "economic problem" then, today we have new *problems*, such as the "body problem," since it can be equally argued humanity has been trying to solve the problem of its own body for millennia, such as doing away with sickness and exhaustion. What are we to do with the plethora of activities we have created to deal with those when they are no longer relevant, such as meditation, exercise or food? There is also the "culture problem," in which so much of human activity has been aimed at adding texture to the landscape and realities it lives in, via art and music, and so what are we to do when these tasks are automated, as well? As we can see, we are not simply talking about the freeing up of time in our day—as was Keynes' anxiety—but of active participation in the essential ingredients of our lives. Since he was worried about a "nervous breakdown" of just a slice of our life given to automation, are we in the midst of an en masse nervous breakdown?

Of course, the world could do much better with more leisure. There is nothing wrong with leisure, idleness and boredom. In fact, idle time is creatively productive. Da Vinci once counseled young minds to stare at clouds to stretch their imaginations. Walt Whitman's infamous "Song of Myself," part of a collection that would change the face of American poetry, begins with loafing on the grass and in his lines we can see the dried grass hanging from his crusted lips as he jots down a sentence or two per hour—or was it per day?—or does it matter? Aside from the need to loaf is the ethos preventing us from doing so. In an essay praising idleness, philosopher Bertrand Russell once quipped, "I want to say, in all seriousness, that a great deal of harm is being done in the modern world by belief in the virtuousness of work, and that the road to happiness and prosperity lies in an organized diminution of work."[12]

Russell points out that during WWI, it was observed a successful economy could be supported by a severely decreased workforce (as many of the men were at war), to the point it was not necessary, when the war ended, for the reintegration of working men and women back into the labor economy.[13]

Henceforth, based on economic performance with such little labor power during WWI, four hours a day could suffice going forward, according to Russell. So, if our working day could be done by lunch and the economy would not suffer, why are we working more than that?

So, what happened? A "toxic" work ethos—work as virtuous rather than a duty—crept back into society.[14] For Russell, the origin of this ethic comes from the ruling class, who have promoted the slogan of virtuous work so as to justify their leisured privilege. While belief in this soft, upper-crust (subconscious) conspiracy is difficult to believe wholesale, what about the work ethos of our time? What about the work ethos of companies today? While we should not put too much weight on the following observation, it cannot be denied that the current image of a healthy work-life balance for creative, successful young people is that of *no balance at all*. The mythos of the tech startup's headquarters filled with pajama-wearing youths coding day and night—with pauses to party, re: *The Social Network*—and often living in a house (or is it an office?—it is both and it does not matter), is a powerful engine to the marketing of these tech products that, like their origins, are geared towards ceaseless production. If you work in a "hip," loungey office space attempting to erase the sense that, in fact, you are trading your time for money—they feel to me like my granddad wearing a Red Bull T-shirt—then you can thank the "laid-back" style of Silicon Valley employment: work is so effusive our offices are beginning to look like home.

Regarding the labor-saving promise of automation, Nicolas Carr, a researcher focusing on the interface of humanity and technology, is right to call our attention to the "substitution myth" operative here, wherein a labor-saving device does not just save labor, but "alters the character of the entire task, including the roles, attitudes, and skills of the people taking part."[15] Not only are our attitudes and skills altered by automation *during* automation—getting bored while your car drives itself, for instance—but your *attitudes* and *skills* outside of automation are altered as well.

Let us give that attitude a name—*automation ethos*. We know it is born of technological and industrial automation, but is now applied well outside these areas to cultural and personal sectors, a "technology creep" in which we cannot afford to assume it functions the same, much as an animal in a non-native habitat is bound to collect new habits, altering not just itself but also the new ecosystem.

Automating an assembly line at work to filter for different-sized washers and place them in bins without a human touch is one thing, but to automate emotional affectation in the form of on-demand software engines that call depressed patients to "listen" to them and talk in a computer-generated voice is another. While assembly line automation does have emotional consequences, a worker feels devalued, a local economy suffers and domestic abuse increases (that is an actual trend in crime statistics), its assumptions are based on efficiency. Automation had to appear reasonable and profitable for businesses to adopt and invest in the new machinery. Automating an assembly line and wiping your mind of bad memories, automatically, as soon as they embed themselves in your neural network (which is nearly possible but coming extremely soon—the *happiness algorithm*) is not part of the same fantasy. And yet, while it would be tempting simply to call contemporary forms of automation something different and give it some analytical space all to its own because of this history, as much as film theory must be distinguished from stage history, the former born of the latter, we should not lose sight of history since the cord of influence of the latter can stretch, or recoil, for centuries.

Contemporary society is now infused with an *automation ethos* or, more specifically, a widely shared physiological attraction to what automation *does*. For an individual to automate something now, it does not have to pass the economic test (efficiency or practicality), but rather it *is a value*, an ethos, part of our perception of a "good," part of what excites us as a people and a thing exciting us. This doesn't mean automation is not efficient and practical, as it often is, but rather since it is no longer constricted to the economic sector, it does not have to jump through the economic hoops. Friedrich Nietzsche understood a value to be that which we affirm as life, as life-giving, and so in what we value we can find what we think brings life. Value and vitality, therefore, indubitably share the same mattress.

In addition to the moving picture galleries, two brief innovations, and there are hundreds, suffice to highlight the automation ethos. First is the much hyped pill that can simulate the physiological effects of exercise without exercise, and second is a device (Thync) providing electrical currents to the brain to help provide energy, or calm, during targeted times. Dean Karnazes, a distance runner and aka the Marathon Man, uses Thync for mental morale when running, since the troughs of mental stimulation diminish far into the long races.

As regards the exercise pill, and as an accomplished athlete myself, what I can say is that the greatest benefit of exercise is to be found in exercising. If you only exercise to be healthy, your relationship with physical activity will be intermittent at best. Yes, to have a healthy body is something, but the painter does not paint to possess another painting—it is *in painting* where the joy is. The same with running—the feelings of accomplishment you get from pushing through the mental barriers far outweigh having muscles able to run 50 miles. In fact, theologians have encountered this problem when they ponder why God did not just create humans who loved God automatically, humans who could not disobey. As the story goes, God gave them free will so they could disobey. Why? Otherwise, what is the point?—their love would be empty. We have an analogous situation.

Automation is valued in itself, as something requiring no justification, but it differs in that it is neither a personality trait nor an object, but a manner of having a process completed. The veneer of mastery automation gives to a product or process, or the lazy pleasure it grants when we do not have to pay attention to where we are going, are immediate seductions of its power, and can explain minor successes, but the economic scale at which it is being developed requires a different analysis. The automation ethos is now functioning in our emotional economies as much as it is in our financial and business economies, and while it is easy to get mired in trying to define what *it is* as an ideological motivator, we should rather focus on what it does.

Automation modifies an organism's activity and yet, while a lot of things modify an organism's activity, automation's genre of modification is altering the process by which acts come to be. *Not having* to apply awareness when driving, interpret the cries of your child, look for a restaurant, find a song on the radio, determine which DNA sequence will have the greatest likelihood of fighting cancer, mull over how someone's passions and sense of humor will match with yours, set your thermostat, decide your diet and so on, all of these now or soon-to-be tasks traditionally required physiological effort, which means calories burned from biological motion and brain activity. Automation is, therefore, decreasing our brain activity and our metabolism.

Now, let us draw some early conclusions to be teased out in more detail later—given the widespread embrace of automation via algorithms, we are creating the *possibility*, or *potentiality*, to slow down the physiological workings of an organism (I say possibility but surely we are becoming more sedate as a species). It is a hibernation of sorts. Again, we think we are automating so we can do more, but the inverse is occurring—we are doing less with less amounts of awareness and brain activity.

Does the Balloon Rise?

Our minds are balloons submerged in water. The hands holding down the balloons are slowly losing their grip and yet the balloons remain submerged. Existential or vitalistic limbo? Or, an alternate ending, we thought the hands were getting tired, but backup was called in and a fresh pair of arms is now in place to hold down the balloon (rather than increasing leisure, we work more). The history of the promise of automation is akin to philosophical marketing, but just like when you get home and unwrap the product, buyer's remorse sets in.

Automation promised so much and then, after not delivering, it found itself implicated in non-native places, such as our emotional and personal lives. Currently, it is a civilizational experiment. The *automation ethos* is a value, since it is redefining basic principles of how we interact and engage with each other and our world. Implicated in what it means to live, automation promises "life," *more life* perhaps, it is just that we have yet to figure out what exactly it is promising.

Chapter 3

Unlikely Extinctions

"This book [the book of the universe] is written in the mathematical language..."
—Galileo

Galileo got a little too excited, right? I mean, we know a ball will drop when we let it go. It will fall. We can even predict its acceleration, velocity and force of impact. We probably know what we will do at work today, roughly speaking, and how the night will go, but what about who will call us today with a crisis or a tragic accident? Can we predict that something beautiful will catch our eye?

Knowledge is a funny thing, often associated with having information in one's working memory (the expert), the ability to calculate (intelligence), or solve, problems. Then there is wisdom, which, while different from knowledge, we agree a wise person must have wisdom, but that someone with a lot of knowledge is not necessarily wise. We have "smart," too? Is John Nash the paragon of smart, the eclectic pattern-hunting character of *A Beautiful Mind*, or is it the fast and loose brain of Will Hunting in *Good Will Hunting*? But isn't "smart" used in reference to smart phones, or smart devices?

Thanks to software, what goes for smart, and conceptions of intelligence, is undergoing alteration, much as it did throughout all of history. But here's the catch—in order for our devices to be smart, our world has to change for the device. Data has to be extracted from our world and put into words, pixels or code for algorithms to sort, and it is, like art, an abstraction. Data is *unreal*. The question is—is it representing the world accurately?

Life Gone Interactive

Scientists and mathematicians have long been seduced by the concept of a mathematical universe, and of a universe that opens itself up to calculation, and hence, the human mind. In 1623, at the age of 59 and in the now famous *The Assayer*, Galileo made the following statement about the great book of the universe: "This book is written in the mathematical language, and the symbols are triangles, circles and other geometrical figures, without whose help it is impossible to comprehend a single word of it; without which one wanders in vain through a dark labyrinth."1 In fact, legibility of the world was, during medieval times, discussed at length in the "book of nature" debate, a debate Galileo inherited and one we are still struggling with. Was there a book? Did God write two books—one about the human heart, the other about His "creatures"? Could we be missing a few pages? Was God a tad tipsy when He penned the early twenty-first century? But even the "book of nature" set of questions goes to the heart of how the human mind is situated vis-à-vis the world, a question as old as cognition itself. Pythagoras, Greece then Southern Italy between circa 570–490 BC, promoted the notion of a unified, ordered universe, to which human purpose was found in combining numerological knowledge with moral and ethical insight so as to achieve a state of harmony.2 In many ways, dataism is a neo-Pythagorean enterprise—the belief that big data, once it has the power, speed and hardware to run on, can harmonize the universe and ourselves in it.

Let's not alienate the sciences—artists and visionaries and religious figures have as well, for millennia, professed some type of direct contact with the stuff, or personas, of the universe. While the jury is still out as to whether or not we can know any *one thing* fully and comprehensively—*an infinite task, no?*—it seems to belie common sense that life, the Earth and everything in it can be known, fully and transparently, and thereby be put into a database as so many techno-enthusiasts dream, with the ultimate goal of configuring the structure of existence—brains, bodies, genes, memories, time, consciousness—in the manner we edit a PowerPoint. This is, basically, what many envision.

Let us revisit Cormen's definition about algorithms: "*Computer algorithms solve computational problems.*" A serious issue presents itself when you think about this definition in combination with the increasing dominance of algorithms in *our* lives—most of our problems are *not* computational. And they are rarely *problems*. Getting home from work is not a problem, it is a task. A task can turn into a problem, but not necessarily. Tasks are acts, sometimes instinctual, but mostly habitual, like putting on socks or driving to work. Completing tasks takes up the majority of our day. In contrast, a problem is something we have difficulty with, something causing consternation. A problem requires a solution. When something is labeled as a problem, it has the specter of negative emotions. In this sense, getting home from work is not a problem, but *getting home extremely fast* is. Simply getting home is just part of life, like so many other aspects of our existence, like preparing food, and while it can be dull or extremely thrilling, experience is not *problematic* in the literal (or existential) sense.

No doubt the draw of automation is the efficiency-infused mindset, in part the product of late capitalist information economies. Constant production, maximum output, never rest, always working, streamlined efficiency, etc. It is an *economic mindset*—time as money—gone awry and our devices are made to enable this. Nietzsche made a mid-nineteenth century observation about behavior that is even more relevant today: "One thinks with a watch in one's hand even as one eats one's lunch whilst reading the latest news of the stock market."[3] Russell echoed Nietzsche in 1932, when he quipped, "The modern man thinks that everything ought to be done for the sake of something else, and never for its own sake."[4] We are talking about much more than being distracted, or being forced to multitask out of necessity, but rather a baseline shift in how we are viewing the *acts of life*, the same acts constituting the majority of our existences for thousands of years. This is the kernel of the automation ethos.

Given that automation is not, and never was, the ultimate goal of technology, it seems rather a case of a *solution looking for a problem*, with the problem being that the applicability of algorithms has no boundaries. The task that formerly *was*— driving, finding a movie, shopping for one's food, narrowing down one's potential mates, remembering something—is now a problem in need of a solution which, as the substitution myth explains, changes the nature of the task. With automated solutions, we look for romance differently, experience riding in a car differently, look at the night sky differently.

One unforeseen effect of the substitution myth is that we are, in essence, allowing experience to be branded. Algorithms are inventions, sometimes they are a company's main innovation, and their deployment in our lives is their product which, when our lives increasingly start to conform to their product, we have branded experience; this type of existential branding might seem to be nothing new to a society saturated with brands and labels and lifestyle marketing, and yet, it is new because it is a brand insertion *prior* to consumer choice. It is a way of monetizing private experience, not just our possessions, and if our era is, as many have claimed, one of the erosion of private space, then this further chips away at what little parcel of wilderness we had all to ourselves. Of course, many of the acts of life are routine, dutiful and outright boring, and one cannot blame the majority of people for wanting to automate it. I mean, when you have Nest, who wants to have to set the thermostat? The issue is, in part, one of attention—when we automate a task we have the ability to substitute one focus stream for another.

Though automation is but one result of the information economy, it does magnify the effects of mass information. In a now famous observation dating to 1971, political scientist and economist Herbert Simon made the following correlation between information and attention:

> ...in an information-rich world, the wealth of information means a dearth of something else: a scarcity of whatever it is that information consumes. What information consumes is rather obvious: it consumes the attention of its recipients. Hence a wealth of information creates a poverty of attention and a need to allocate that attention efficiently among the overabundance of information sources that might consume it.[5]

We all know everyone is distracted today and that there are just too many blogs, stories, viral things and amusing tweets, and too much tagging, linking and commenting to do the rest of life justice. However, the latter part of Simon's observation—that of needing some way to allocate attention—could be an early argument for the mass introduction of algorithms into personal and professional lives; for example, algorithms help us to separate the wheat from the chaff, so we can focus on more important things. The problem is, of course, our attention goes to something equally pointless. This means that nothing is ever "solved" when it is automated, at least regarding our personal and social lives.

Algorithm-Ready Experience

A solution is really only a solution when it is computational, that is, when it has a definite answer. The issue: life does not provide definitive answers and we should entertain no notion that it does. Despite the wishes of some philosophers, engineers and scientists, we are not rational creatures. And yet, if automation is to work within the intelligent systems of the future, which it will be, it will have to be put into a rational system. More specifically, according to one expert, "experience has to be devised as a mathematical representation."[6]

In order for experience to be devised as a mathematical representation, it has to be rendered and put into another "format." For instance, Google's self-driving car does not "see": it computes inputs. Another car in front of it is not another car, with a potentially drunk driver or a distracted teenager with Coldplay bumper stickers, but a set of data: speed, velocity, shapes, width, tonalities, trajectories, tendencies, etc. This data is mathematical in nature and, since this data must be able to put into conversation with one another—the car's AI must be able to weigh a car's speed against its distance, for example— the data must be modeled and tagged accordingly. In short, the experience of a car in front of it is broken down into bits of data needing to be compared, analyzed, sorted, filtered and so on. Obtaining a data set from experience as we feel it, therefore, requires an abstraction. Hence, the oft-quoted maxim that "there is no such thing as raw data."

Likewise, Modern art is equally an abstraction, a second-order representation existing alongside life, some would say incorrectly "above." So is poetry and so is realistic figure drawing or photography...an abstraction. Life, we assume, is something immediate to us, something that does not have to pass through a filter, and yet, this is incorrect, since sensations and shapes and sounds become such after passing through our brains. As the infrared spectrum highlights, red is a frequency, and there are many other types of sounds and colors that we cannot experience, given our organism's limitations. In other words, there is no such thing as raw experience, just as there is no such thing as raw data.[7] Data is poetic, in this sense, but there is a qualitative difference between the data of poetry and the data of data—the latter is arguably the most powerful force in our world today, and this force is rational. The second-order representation of data must be in a logical structure or set, which cannot be said for other representational orders. This begs our initial question of—*is the universe rational?*—to which we could answer provisionally, it is going to be if we let it.

But about us—if the current state of global affairs is any indicator, we are a passionate animal, worried about hunger, safety, our children, things we love, the future, pets, our past, pleasure. We can be rational, but we are rarely content with rational solutions. Sacrifice, forgiveness, joy, fear, murder—these are not rational. We are emotional beings with rational outbursts, at best, and while we pursue science in a rational world seemingly governed by rational behavior, we are getting left behind—we still seem to be the fickle ape blowing on a precious ember.

The Extinction of Taste

One of the problems in writing about the costs of automation—for example, uselessness—is studies are hard to come by. It is one thing to track the number of laid-off workers in Chinese manufacturing who lost their jobs to robots, but it is another to track how Pandora Music & Radio, Netflix and Waze—to name the big ones—are undermining our ability to be guided willfully, vital human beings, to be discerning bodies of taste and culture. The problem is largely one of figuring out what a scrap of culture looks like, since if we can do that, then we are able to see if said scraps are being degraded—like human willpower or instinct. In their research on cultural transmission, Jelmer Eerkens and Carl Lipo state the problem succinctly: "Unlike genetic transmission, there are no agreed-on empirical units of cultural transmission."[8]

It is one thing to track the disadvantageous effects of a lack of mindfulness in an automated world, as Nicolas Carr and others have done, but it is even more difficult to track how automation is bringing the very *idea of taste*—as a personal aesthetic for an object or event—close to extinction. In other words, when movies are presented to you, the music just streams and an app tells you how to get home, you have lost a critical moment when you are forced to make a decision on the things you find beautiful or pleasing or, for that matter, wretched. In this loss, you have therefore atrophied a piece of willful self-definition (agency) and, with that atrophy, rendered your tools for cultural creation dull, ineffective and useless.

Pandora makes associations about the music you have already listened to and what others also listened to, but you didn't seek it out. You never really trained your ear to cultivate a personal sense of beauty and that is dangerous, because a personal sense of beauty is vital to individuals and the heartbeat of societies. If we can partly agree that the experience of beauty is one of the most cherished parts of living, oft cited as the point of life itself, then we can agree that losing our ability to cultivate it is ill-advised, as dangerous as allowing a leaking oil tanker to park in our backyard, directly above our water well. In the early formation of the Internet it was expected the sheer volume of stuff on the Web would be a great experiment in exposing us to cultural differences—I mean, you can in two clicks expand your world and witness Nigerian tribal music and Arab feminist discourse and Mongolian camel riders—but the inverse has occurred.

Amid the panoply of global output, we have become narrow in our choices and our exposure becomes myopic, a self-reinforcing loop. This self-reinforcement loop, whether we seek out what agrees with us, or whether it finds us automatically, arrives via algorithms. An example of this danger comes from German Chancellor Angela Merkel, when in 2016 she warned that some social media and news algorithms—Google and Facebook—are politically dangerous.[9] Merkel demanded that such algorithms be made public. But this should come as no surprise, as Alexis Madrigal has succinctly noted, algorithm's *raison d'être* is to restrict information.[10]

The informational myopia works on us spherically, in a *push* and *pull* phenomenon. We pull content toward us that resonates and content is pushed at us from the algorithms analyzing our clickstream data, user habits, demographics, predictions and so on. Likewise, content the algorithms deem prickly to us does not show up on our feeds. Eli Pariser refers to this as a "you loop," which is essentially a you-affirming data set catering to your idiosyncrasies in real time. Despite the fact that oftentimes these digital matches, such as ads on your newsfeed, miss the mark completely, the ability to predict what you actually want will get only better and pre-emptive purchasing will be a thing in the future.[11]

The fact that the Internet of Things was cooked up and slated to be the thing of the future—since there were a thousand other applications to new technologies—was initially germinated in how Web companies thought we wanted to be catered to. Which isn't the same as how we *should* be catered to. If you visualized this habit as that of someone you are courting, be it a boyfriend or girlfriend, and their behavior was constant affirmation, constant appeasement, never wanting you to struggle, always excessively trying to "be there for you," never wanting you to leave—we call that overbearing, weird, unnatural, insecure or just plain stalking. While the "trending" items might be ubiquitous for all users, other "recommendations" are not, the latter based on carefully crafted and trademarked algorithms designed to reflect, inform and shape your data, since we are not so much consumers of most Web products, but products (sets of data) they sell. Web users—for example, our data—are the product.

Netflix gives you a range of options now on most streaming services, but it only provides a fraction of the films it has on file. With traffic apps geared only toward efficiency, we thereby take the fastest route home, not the most beautiful. In all instances, we fail to carve out an event or thing that is ours. French thinker Michel De Certeau remarked how human walking habits instinctively resist the grid of city streets, opting for circuitous routes along tree-lined boulevards, or statues affording pleasurable pauses; often, this walking inefficiency is unconscious, but it comes from us and is responding to something beautiful or pleasing (releasing tension), however mysterious the former.[12]

∞

Algorithms provide solutions to problems; outputs from inputs. The issue is—life, and the tasks we perform in it, is not problematic. However, when automation becomes a value (*automation ethos*) we think something ought to be automated for the sake of something else. Automation is basically a promise of time, a *lifestyle product* without a material form (other than code), since it is a substitution. In this milieu, life *does* become problematic; for example, filled with tasks we would rather not tend to. Yet, a solution is never really offered, but merely deferred. If we really spent our time more wisely, then automated products are successful. But we don't. Some automated products do make offers, such as films or songs or food, but what is really offered is disengagement, the forerunner of uselessness.

When lack of engagement reaches a tipping point, taste, as an attunement and comportment toward the world, loses its critical teeth. Culture becomes unmoored, its oars out of the water. Culture is transmitted in a myriad of ways, but no one agrees on its units of transmission. We know culture is transmitted from parent to child, child to parent, through film, shared experience, social media, images, likes and memories, yet increasingly it is being transmitted *without* our intervention. As a result, the inclination to become willfully guided agents in our environment is becoming increasingly irrelevant, in other words, useless.

Chapter 4

Up the Food Chain

Why is it important we decide on anything in life? Why cultivate a sense of taste? Who cares, really?

Taste—it can go by other terms: choice, preference, things we like, buy, sleep with, eventually love—is the amorphous arbiter of the stuff of life, the things we surround ourselves with, both on a microscopic scale (the type of books we have on our shelves) and a macro scale (the design of our office lobbies, walkways and storefronts). It is a mistake to moor the concept of taste to consumer preference, i.e., the phenomenon of taste is the product of late capitalist (and individualist) societies.

Taste can be shared and it is universal. No culture to date has been agnostic about how it surrounds itself, just as for an animal any old nest will not do. Taste is an outward-facing vote, an act of our will we place onto the world. It is equally about creating things, cultivating things, as much as it is consuming them. Without human taste and the trained eye of discernment you have brute materialist pragmatism, a disengaged public and questionable beauty; you have, in short, a type of anti-civilization; even those trying to "destroy" culture, such as when in 2001 the Taliban blew up Buddhist sculptures, which stood proud for 1700 years, they were trying to establish a *new culture*. Eventually, once it accumulates *up the food chain*, taste creates the flavor of our world's surfaces and edges. Taste is a decision, nearly imperceptible, equal parts rationality, instinct and gut, by which we transmit the things of the past and present to the time yet to come; in each instance of taste, therefore, is a vote of confidence. It is a decidedly philosophical phenomenon. Taste is a shovel that tosses the dirt.

By refusing to vote on the things we consume, or at least

stepping out of the way, we are building an economy with no emotional driver. The assumption here is that after so many objects are put in front of you that "suffice" — Netflix's recommendations are not too bad, by the way — you get a culture and existence that suffices. On the other hand, it cannot be denied that civilizations have always relied on mediators of culture, and that the *New York Times* in the nineteenth century deciding what is important to read in science, business and art is not much different from algorithms guiding knowledge consumption in the twenty-first century.[1] Whether we are talking about elites, cultural creatives, seers or gurus, the phenomenon of something filtering out the noise of culture for the rest of humanity is not really a phenomenon, but substantiated by the vagaries of history.

But the scale is now different. Whereas traditional cultural monopolies might have restricted themselves to industries — such as film's Warner Bros. or J. Crew'ish fashion lines — rarely, at least in "free" societies, has there been the same operating principle working underneath various industries. If, according to nineteenth-century literary critic Matthew Arnold, the purpose of culture is to determine "the best which has been thought and said," the wheels of culture are clearly trending to malfunction.[2] Whereas the traditional curators of culture — whoever they be — at least made an attempt to retain cultural vitality (yes, this is a dated and romantic view of cultural elites) we now, or will have, algorithms dictating the fundamental direction of culture, algorithms that cannot have vision for humanity by virtue of not being grounded in the joys, pains and failures of human life. It is in this manner that we live in an "algorithmic culture."[3]

Algorithms are not just offering a platter of creative objects with its own brand of creativity (Apple's "Genius" function simply creates playlists), but they are *playing the artist as well*. Under the tutelage of market pressure, software is making decisions on how the most important myths of our time are being written. I am talking about film. Whereas myths used to be shared narratives about what makes us tick as human beings, passed on from the elders to the young in rituals and generated from the backs (often sweat) of human suffering, now we have software editing our myths; yes, independent films are going to continue to get made, but that really is not a counterargument. The threat is elusive but real.

Nothing can replace the passion and emotional experience a human mind adds to the creation of writing characters, or relationships, and when we remove the human from the writing process we, as consumers, remove one more opportunity to grow, feel and work through our own issues. But, expectedly, there are researchers already at work training algorithms to detect the emotional "shape" of a work of fiction, with the ultimate, if largely unconscious, goal of teaching software to write "emotional" works. By analyzing how certain words are associated (in fiction) with emotions, the software was able to predict with 50 percent accuracy the genre of a novel. No big deal, right? Well, of course, once the software knows how a great work of fiction distributes emotional words, writers could upload their novels to the program to determine a more optimal emotional arc (movie production houses already practice a primitive form of this scanning).[4]

Unsurprisingly, this debate is not far from the "value of the arts" issue which, since the first instance when people thought about the strange phenomenon of art, has been difficult to answer. This has not stopped many from trying, including noted art's advocate writer Elliot Eisner, when he observes: "The arts enable us to have experience we can have from no other source and through such experience to discover the range and variety of what we are capable of feeling."[5] At the very least, creative objects provide creative experiences, which should not always be considered positive—art challenges, hurts us, makes us feel uncomfortable. What newsfeed is going to try to offend you in that positive, "glad I read that" type of way? Art and the best of culture is like great comedy—it exhumes the assumptions paining us. Disagreement is a critical tool of our physiological development and if we remove the stone on which to sharpen our teeth, then we have in effect not just developed in a certain direction, but not developed at all.

Future culture will be unreflective—in terms of engendering reflection in its people—since the embedded shrapnel of human introspection, the fiercest power we have, is lacking in stuff not created by us. The Copernican revolution was a violent idea for its time—it put the sun at the center of our solar system, shoving aside the notion that our planet was the center. Analogously, the automation revolution is doing the same for human agency; while the Copernican revolution was a net boon to civilizational intelligence, the automation revolution—while likewise marketing itself as a boon to intelligence, and as a result thereof—is a loss to those deploying it.

A decent, and popular, argument in favor of algorithms is that they are neutral; since it is *not* a human selecting what is best, then the process is not *tainted* by what history has often revealed—cultural drivers or those in power often promote their own agenda, which can be racist, misogynist and so on. The myth is that your algorithm-fed choices—such as Google News—are formulaic, neutral, objective and do not concern themselves with unrelated issues, such as politics. But this is not so. Not only are there tremendous amounts of human labor and maintenance that go into a functional algorithm, like a Netflix recommendation or Twitter trend, what goes into them is often hidden behind corporate copyright and intellectual property laws, hence why Merkel demanded that algorithms be made public.[6]

An algorithm is far from neutral or objective. As Tarleton Gillespie has argued, algorithms are not proxies for the "web-wide collection opinion," but follow a carefully crafted logic built on crafted decisions by companies, executives and political nuance.[7] What we know for sure is that we do not know what goes into an algorithm, which makes our (blossoming) algorithmic culture a set of clothes with no emperor, guided by an invisible hand.

Gone Life Interactive

Since algorithms are now writing business and sports articles, and being "apprenticed" as creative writers, what is to stop them from producing full-scale script development? It is nearly impossible to quantify how the loss of the arts will affect our emotional lives, but common sense easily provides an answer— we will become more useless, boring, less interesting, and the mental health of our progeny will be at risk. Is this the origin of the apathetic anti-hero of so many indie films, who can't make a decision, feels distanced, unengaged?

The Internet of things, premised on the idea of all things turning toward us, is quite seductive. It sounds like *effortless life* and, since we are all exhausted from scatting about a world at warp speed, it is a welcome opportunity to outsource a few tasks— all things bending toward us, set to just how we like them. But, really, is life that uncomfortable? Who wants an environment you do not have to interact with, you do not have to work for? It would surprise many of us that manual labor and using your body is the foundation of not just a few religious orders, such as Christian and Zen monasteries (after enlightenment, "chop wood, carry water"). Effortless life is, in many ways, less life.

Tech IOT developers actually pitch their products as *life gone interactive*, but it is actually the opposite. We interact with life, our environment, much less with this cutting-edge technology, and studies on boredom, as we shall see, consider engagement and interaction a key principle to avoiding boredom. The more we continue on the path of having rooms and houses and public spaces cater to us the more dull our tools become to do that interaction, and our expectations lower as well, in a type of positive feedback loop affecting us negatively—situational awareness, adaptability and humility.

Daily life is made of a thousand little decisions and the fabric of our lives—our relationships, homes and beliefs—is the product (material or otherwise) of those decisions. When you lack the opportunity to make those decisions, you become complacent. When you go with the *digital flow* (no, automation is not turning us into Daoists), your input into the collections of habits, rituals and things that make society is lost. We participate in a society in as much as it is symbolic, in the sense that we are participating in a living history made by living beings. You are no longer a contributor to the layer of living history, and meaning itself, when you "check out" from all the goings-on in society. When you stand back and take stock of all the things we have automated for us, the things that are automated without us knowing it, and the things to be automated, you get a staggering proportion of cultural objects and participation at events lost. Eventually, not only are you, literally and figuratively, useless, but the society in which you are a part becomes useless as well.

Part II

Culture

Chapter 5

Is Technology Creative?

"The best minds of my generation are thinking about how to make people click ads...That sucks."
—Jeff Hammerbacher, an early data-analyst at Facebook

Adventurist, poet, pilot and writer Antoine de Saint-Exupéry once said, "If you want to build a ship, don't drum up people to collect wood and don't assign them tasks and work, but rather teach them to long for the endless immensity of the sea."[1] While this seems like a biblical proverb dressed in new garb—the whole "don't give a man a fish..."—it is not. It is about desire, not a set of skills, about motivating the heart of a people, not technical knowledge. And we should not confuse the two.

Let us bring Saint-Exupéry up to speed, and substitute some key words—"If you want to build an app, don't drum up people to teach them *code* and don't assign them tasks and work, but rather teach them to long for the..." It is precisely the "..." in question, and it is this ellipsis fueling the race to leave no process unautomated and causing venture capitalists to refer to automation as the new "gold rush" for investment. What we should not forget is that apps and IOT devices are commodities we purchase; they are entities businesses sell—more often than not the "startup"—and, if there is one rule to observe about commodities in the "society of the spectacle," it is that they promise without ever being able to fulfill said promise. Unfulfillment is what fuels late capitalist "lifestyle" products. But the pleasure we get from Pandora or Waze or Siri or Tasker is a strange pleasure, not quite a traditional commodity. We consume them nonetheless. So, the question is—are these objects creative? Are they making us more creative? What is the nature of these things?

The launch of a new automated product (in the form of an app or program or device) is becoming a "new" type of creative act. A startup's initial Kickstarter campaign or their IPO with the CEO in tight jeans and surf hair is a modern form of an art gallery opening. More children today can cite the names of tech CEOs than at any other time in recent history. Just as the "painting" is a place where a creative mind can be exercised, where young and old flock to find themselves, so, too, is automation a place trending for creative minds, and Silicon Valley currently is for tech what the East Village was for poets in the 1960s.

Tom Campbell, a writer for BOP consulting, which specializes in research and trends in cultural and creative industries, recently warned of the automation death knell for the so-called creative class—designers, writers and artists. And he seems to be OK with it. Campbell wonders aloud why we are all somewhat complacent about industrial jobs getting automated—"well, yes, a robot will do just fine to rivet all the bolts on a car." And then Campbell adds, "Much is often made of the 'creative process,' but it is just that, a process, and as such, it can also be an algorithm."[2] Is his statement accurate—can the creative process be turned into an algorithm?

As it turns out, the mechanization of art or artistic process is not a new topic, at least for those who know the eclectic debates about art theory and practice. Andy Warhol, the most famous artist for pushing forward a mechanistic art process, said as much in a 1963 interview with Gene Swenson of *Art News*: "The reason I'm painting this way is that I want to be a machine, and I feel that whatever I do and do machine-like is what I want to do."[3] Warhol didn't want to be a machine, but he was pushing against the notion of the heroic artist's struggle of wresting the idea from the soul via turmoil, you know, the suffering servant artist whose agony only sweetens the reward of creation. It's an old myth. In other words, he wanted to take the metaphysics out of artistic generation—the more automatic the better. But Warhol is just a good place to start and art has been impersonal for decades. Michelangelo had troves of helpers, same with Rodin, and many high-producing artists today run their studios like CEOs, directing and monitoring the creative process. Quasi automation.

So, if one can still be an artist and not really hold the paintbrush, what really defines the creative process? Christopher Steiner, a TED Talks veteran and author on algorithms, writes, "In many ways, creativity means incrementally building on past innovations, and there's no reason a machine could not do that."[4] Steiner writes about programmer and professor David Cope, who has created a program able to compose classical music so well that many, including musicians, cannot tell the difference. There are also music algorithms able to compare your song to the hit songs of the past—just upload your MP4 to a website, of course, and well, if your song's data points closely resemble those of past golden hits, you may actually have something.[5] Music producers are already using the software as an early screening device, just as companies are using résumé scanners to filter résumés—looking for keywords, salary tags, verbs of accomplishment—before they get to a human set of eyes. Wouldn't it be much cheaper just to automate the whole thing— from song creation, production and distribution—without ever passing the human judgment test? Imagine the profit margins on a global hit if a software algorithm could write and produce and distribute a song in half a second. Imagine what artists could now do, given their newfound free time. The submerged balloon rises.

Origin Matters

Devotees of creative algorithms forget one important thing—our species values truth over deception. Actually, this is completely untrue, we probably value deception and illusion over everything, but I think the point remains—the provenance of a creative entity, such as the "emotional grist" of a novel, matters. Origin, so indicative of truth, matters. Imagine you are engrossed in a novel about a pregnant young girl fighting for her independence and you are glued to the pages for a week. You cried, told your daughter to read it, lost sleep. You wondered how on earth the author penned such a thing. My bet is you would be offended if you learned this novel was written by a computer program using characters from trending books, tweaking the storyline based on user data, changing names and places to maximize readership, and "composing" the book in nanoseconds, as long as it takes a calculator to crunch numbers. Furthermore, it will be possible in a decade or so to have a novel written *for you*, tailored to your tastes, personal history and behavior objectives. IOT for happiness. You will be able to buy your custom novel off the future's version of Kindle, if by that time we will even need to read the old-fashioned way, via our eyes moving across letters. With memory implantation and deletion services already on their way, why not just have the contents of your novel uploaded automatically to your memory bank?

Origin matters for another reason—works of art are not just entertainment wherein the user should be oblivious to the *connectivity value* of the work of art. Paintings, sculptures, even good tweets are, and should be, empathic products: causing us to be aware of other experiences that produced the work of art. An aesthetic experience starts at a thing, but expands outward. The life-sweat behind the thing responsible for its life is what animates us, and it.

While algorithms are more gifted than us at recognizing patterns, the structural assumption of the algorithm is *past-facing* and the type of creativity it engenders is based upon what has come before it. Since algorithms need inputs before they generate outputs, their models of prediction—such as the new hit pop song—have to be based on past pop songs. Steiner is absolutely correct to note creativity builds off "past innovations" and resultantly, most creative moments require a repetition with a difference. That is, the genius of Michelangelo would never have existed were it not for Giotto and Giotto's genius would have never flowered were it not for the *christus patiens*, iconoclasm, Jesus himself, and so on into the annals history.

The "difference" in "repetition with a difference," however, is not an algorithmic difference, but a qualitative difference born of a human struggle, an individual who has breathed our air and speaks, witnesses, textured life. How can a software bot running on computer code understand the future to have intrinsic human value? True, a lesser artist may create something with algorithmic difference, but in that case the artist rarely makes history. Sure, art can be mechanical, but not great art, and this is because great art is at once historical, human and cultural. Great art is never limited to the product, nor the past, nor a redistribution of the sensible. Even while some claim algorithms today can create in forward-facing manners, and do not have to take their cues from what has come before them — true algorithmic novelty — we have to ask how these "cues" are implemented. Are they simply random? Human-consulted cues? While it is also true not all art has to have a humanistic or artistic purpose, it nonetheless has one, given its provenance. To say otherwise is to claim the flower has no soil in its constitution.

A growing divide between art and humanity is inevitable if history cannot be accounted for in the production of art. If more and more art is judged on the basis to other art alone, then creativity lives in a vacuum and, indeed, the logical result would be a real-time biometrical pairing of taste and automated artistic production, where songs, film, visual art just begin to appear around you in various guises as you live, an art as seamless as it is insignificant. Neon Labs, whose research has its roots in Carnegie Mellon and Brown, has claimed to have discovered a "computational model," read algorithm, which can select the most engaging images for public consumption. "We collect large amounts of images," said Sophie Lebrecht, their chief executive, "[Then] we run them through a computational model for how the human brain sees and responds to images and that allows us to show the types of images that are going to drive engagement."[6]

In other words, software is already selecting what goes for beautiful...but not for the sake of beauty. The purpose is to get us to click on ads, videos, stories, what have you. The intersection of math and art is not new. The ancient world gave us the golden ratio—which they believed was a secret formula to create affects of harmony, pleasure and beauty in our psyches and bodies; it was used in painting, architecture, urban planning and sculpture. The golden ratio was extensively used by Da Vinci 1500 years later and is still used today, notably by software company PhiMatrix, whose slogan reads: "Want to go from ordinary to extraordinary? Here's a little secret: Design with the same proportions others see as extraordinary."[7]

Eternal Recurrence

Nietzsche, the great German stylist, had a fear that kept him up at night—it was called the eternal recurrence. This was Zarathustra's most dreadful thought, Zarathustra being one of Nietzsche's most infamous characters in *Thus Spoke Zarathustra*, written between 1883 and 1885. The eternal recurrence was the thought that, since there is a finite number of materials and combinations on our Earth, everything was bound to be repeated. Life rendered uncreative. Nietzsche was talking about more than just history repeating itself, but more specifically, given a finite number of ways the world can organize itself, the experiences produced by the world will eventually have to repeat themselves. While there is no discernable way to validate or invalidate the thoughts here, since Nietzsche was speaking as a theoretical philosopher, we can most assuredly say that, given the direction and pace of software intrusions into life, the range of experience is going to be limited concretely.

Before we can ask about the consequences of consuming "creative" tech objects, we should ask what is being copyrighted in automated software? It has no added utility value in the literal sense, since commodities are *additions* to the consumer (or user) experience. Apps utilizing automation are in the business of *removing* an experience from the user. Since meaning is generated in experience (these are at least the general parameters of the meaning-experience debate), the removal of experience renders the object meaningless. This means that some apps, quite literally, have no meaning or intrinsic value other than *their ability to remove meaning*. This is the essence of automated commodities.

A simple but profound reversal has occurred, as colorful as it sounds, and it is this—if the old model of artistic creation was that of *something from nothing* (God making Earth from the void; the artist wresting the idea from his sentient soul), now it is the creation of *nothing from something*. Or, more precisely, it is the silencing of cognitive brainwaves, not for the exercise of our "higher aspirations"—the promise of leisure—but for the thrill (is it a thrill?) of *not* having to deploy our conscious willpower and move forward on the Earth with intentionality. Uselessness. Yes, allowing our software to make decisions for us is a type of deployed willpower, since we do make a decision to let it make decisions, that much cannot be denied. However, it is a choice that, once made, is viral and rhizomic—it makes other choices on its own accord, independently. Then, we forget it is doing so. And yet, computer code is legally considered a work of art, and the US Patent and Trademark Office recognizes it as such.

Not surprisingly, computer code is literally getting auctioned as works of art, in the manner of a Picasso, since, like the latter's *Guernica*, algorithms work on and affix culture in new directions.[7] John MacCormick's singling out of specific algorithms that have changed history, such as "digital signature" and "forward error correction," tracks how early algorithms, most of which you cannot identify but use every day for basic Web services, have already changed your life more than, say, great teachers you have had.[8]

If we can allow ourselves to get caught up in semantics for a bit, what's with the word "innovation"? As it turns out, the crowd distinction between creation and innovation reveals the subtle but important distinctions we all make about science and tech products. While we are often OK to call Steve Jobs an artist, it seems jarring to call American painter Jackson Pollock an innovator. Sure, Pollock did innovate, but he innovated *within* painting. He was a painter, first and foremost. In contrast, Jobs can be thought to create, but the justification of Jobs' creativity lies in the success of his products. If Apple's products all flopped, then he may not be crowned with the title of artist, except among some cult circles; undoubtedly, success is not the determining factor for an artist-CEO, but it is a big factor since it is in the realm of business (in other words, profit) where he works. On the other hand, we seem to be OK with calling an artist an artist even if they slave away in their cramped NY studio apartment and paint their radiator for ten years. In fact, they may be *more the artist* for doing so, for living so recklessly, a quality that should, on my accounting, remain within the artist. Surely, we expect a tech CEO to have wealth, even if, for instance, Zuck still drives a beater...But what are we saying with all these words and distinctions? While not all tech creatives seek wealth, tech products are creative entities within a profit-driven industry.

Tech ideas are for-profit ideas or, at a minimum, this *class of idea* remains lodged within a profit-driven economy, nurtured by big data and the drive for efficiency. As a fellow creative needing to fight the interruption of creative intrusions just to complete a day's work, I understand quite well the artistic impulse to exploit hitherto unexplored terrain. Here is what you are thinking—artists must explore. Artists must not think of consequence. Artists, in order to "listen" to the real voice of the creative process, must be present with the creative moment to such a degree of temporality infinity any intrusion of practical necessity or future practicality must be kept at bay. Like a small bird, if you hold it too tight you might crush it. Just follow the muse and the idea will find an application. In many ways, the startup—bolstered by their coveted "idea," much as a fine artist—is a creative entity, off the ground quick and often quick to die. The startup is feverish, reckless, an insomniac, youthful and daring.

What tech artists could learn from art history is that art must be weary of business unless, of course, it has no problem with its creative objects being associated with profit as opposed to vision, two entities that ought to be mutually exclusive. It must not find itself compromised. There is nothing wrong with making money off art and with this being your sole purpose, but you have to be prepared for the production of bad art.

The. Most. Boring. Future

Art and prophesy. Art and the future. These couplets touch some of the core relationships of creative making. As we saw with Oscar Wilde, it is a common feature of automation prophesy to think the purchasing of more time via apps and software will inevitably lead to more leisure, which will "naturally" lead to the contemplative life; it is like drawing an arrow from automation to art. But what do the inventors of these products envision of life after automation? *What about the future?*

One example of this future can be found in the marketing material for Google's self-driving car (no offense to Google, of course), especially in the "lifestyle" imagery it promotes of what we, as users, will be doing in the car; i.e., what we will be doing with our newfound leisure. We should expect from Google a "lifestyle" revelation, since the dream of fully automated transport is not only found in germinal form thousands of years ago—effortless movement—but also since the invention of the wheel circa Mesopotamia 3500 BC. It is the human dream of ridding ourselves of drudgery and the mindless labor of labor itself. Supporting the dream is the freeing up of our time and resources, so we can pursue nobler goals. Behind great lifestyle inventions should be a bold vision of new life. Da Vinci, for his part, envisioned the "flying machine"—the airplane—to find part of its utility in gathering snow in the mountains and flying it down to the hot city streets so sun-scorched bodies could cool down and throw snowballs in the summer.[9] Now, that is an invention with a creature comfort to match. The high test Renaissance imagination—how to judge the quality of an innovation—it turns out, was as humanistic as its paintings.

So, one might expect the bold-visioned Google—one of the most "avant-garde" of tech companies, which basically means for most people, the most creative company in existence—to provide an equally stimulating vision of the effects of automatic car technology; for example, higher aspirations. What is their vision of what we are to do with their new invention? Make love? Laugh with our kids? Watch the forest as it passes us by? Write poetry or skip work and have the car drive us to the desert as we laze around as if it were Sunday morning? And so, drum roll, what is the image of humanity?—people on laptops.

But there could be reasons for this. First, it could be that Google has expended so much of its creative juices working on the car it forgot to imagine what we are to do when the reality sets in. But let us not be so harsh. After all, they are creative people, creative young women and men driven by the need for invention, for creation. Every artist wants to change the world in some fashion and these "tech artists," with their code and apps as their products, are no different. Still, the inability to provide an equally inspired vision of life accompanying these products seems to be endemic to the industry, to this type of creation: as if it is obvious we need them and that life improves.

As if the innovation speaks for itself—is autotelic—and needs no further justification. But that is a mistake and to fall victim to the automation ethos, which is as untested as it is juvenile.

Apple is extremely adept in lifestyle marketing in non-computational spaces—a forest, for example—and inserting the iPad between the person and the place. Families camping, looking at trees, taking walks—the commercials put screens between people, and between people and their environment; point the device at a tree and the algorithm tells you what type of tree; the same with constellations. While the purpose here is educational, according to research this type of applied knowledge does not settle into our memory banks efficiently because it lacks context. For instance, a little girl will remember a type of tree if her father takes her for a walk in the woods and spends the time teaching her about that tree, but when she has the computer do the work for her, her brain never bothers to memorize it and the neural pathways never materialize; her brain assumes the information is there, waiting anytime she wants it, and so will spend little subconscious energy cataloging it. Further, the memory of the tree along with her father paints a fuller picture of the experience and the memory is therefore more lasting. Memory works on association, and when screens are placed between us and our environments, our worlds become privatized and unassociated. When screens come between us and our landscape we are consuming *the screen*, not the landscape, and we should not reduce the latter to the former. It is precisely this type of reductionism that tech requires, both in its data models and in its algorithmic outputs.

Kurzweil and Rothblatt speak with enthusiasm about how future populations, thanks to advanced technologies, will be able to consume experiences, natural and cultural alike; though, it is hard to see how the former is not folded into the latter. They speak about futuristic *mind clones*, our inevitable digital dopplelgängers to exist in the not-so-distant future, and the mind clone's desire to consume digital art. Mind clones are, in essence, automated forms of human consciousness, copied and downloaded from their original hosts—us. Rothblatt writes, "While the mindclone will be stuck in cyberspace, he or she will still be able to read online books, watch streaming movies, and participate in virtual social networks."[9] Again, if the vision of the futuristic mind clone is for our consciousness to exist on a cloud server of sorts so it can read online books, it seems unlikely that when our minds are fully integrated into a cloud-based data platform—with no bodies nor ears nor eyes—we would want to read an online book when, in theory, the text of the book is already immanent to our brains. It is just there, on the Web, just like our brains, and all we need to do is call it up. When the data of cultural objects is pre-logged or fully immanent to a mind-software hybrid, such as in the mind clone, from where comes the *desire* to consume cultural items? Whence the desire to learn? There is no learning curve, nor the joy of learning, to a mind-bot that knows everything. The world comes to us, we do not go forth into the world, and if the hallmark of fantasy is having the world conform to your wishes, regardless of conscious intentionality, then tech products are not just producing fantastical products, in a sci-fi manner, but also beholden to the fantasy complex, which assumes reality is problematic (data modeling; algorithms as solutions), perhaps owing to trauma, or fear, in need of modification, and that with enough modification a state of pleasure awaits.

As seen in the mind clone example, one among many sketches of future humanity, the intention of automated AI life is not for us to leave the data stream of computation, but to enter it *more fully*. To create products that will lead to us leaving no data trace—to in essence render the product useless—would not be a good business model. A better business model is to render humanity useless, since that translates to greater product usage. Algorithms become more refined the more they are used and the more data passes through their servers. Algorithms become more useful to us the more we use the products relying on them, which means every moment we are not providing inputs and it is spitting out outputs is a *lost opportunity*. To sever the lifeblood of their continuing relevance is not the intention of any business, or organism for that matter—in other words, the intention of AI and automated tech is not for us to sleep under trees, lollygag around the junipers nor contemplate life's mysterious wonders. Sorry Wilde. No vision can be found, because there is no vision of life outside of data production. The rhythms of life, love and joy are given over, in our bedrooms, kitchens and cars, to maximum data output.

∞

So, why is tech unable to provide a vision of the future? The problem is technical, since the version of creativity deployed, and largely put to use, is past-facing, not to mention profit driven. These two factors are not problems in the general sense, but when you have this genre of creativity being deployed at the level of software production and at the level of mass consumption, there is a trickle-down effect at many levels. Only then, when it is affecting how creativity and culture communicate, is it problematic. Another answer could be that the innovations occurring are so myopically idea driven they fail to contemplate the world they are creating, rather just seeking to "change the world," under the assumption any change to the world is worth pursuing. Another answer could be that since one result of automation is the removal of experience, such entrepreneurs do not consider it their duty to replace the gap in time, assuming, as many in history have, humanity will rise like the balloon untethered. Another answer could be that it is because its objects are not *really* about the future, but about letting us live without decision in the present, and since it is in willful presence where we access the past and the future (memory, reflection, calculation, etc.), our future consciousness, if taken down this path, will be unable to think about the future; hence why we are having early difficulty doing it now.

Chapter 6

The Department of Joy

When we try to pick out anything by itself, we find it hitched to everything else in the Universe.
John Muir

You get up. Nothing excites you or depresses you. It is a normal day. More good than bad. You shower, put on your clothes and get to work. No emails to answer—nothing, really, to attend to. Your job? It does not matter. It is a good job, pays the bills, provides for your family, vacations and all the extras...but, it's boring.

What should work be? Should work be fun? What about mildly fun? Mainly dutiful yet rewarding? Should there be no work at all? What if we all just got a paycheck? But, wait, doesn't work make you a good person, build character and all that? If someone pulled you aside and said that in the future 50 percent of jobs are going to be boring, really, really, *reallllly* boring, what would you do?

∞

The future does not bring technological challenges. Technology will advance just fine and it will continue to do mind-blowing things. Technology will not pose the greatest threat to our species in the coming decades; rather, the greatest threat is us-related, with *our* utility. To anchor this thought into a concrete example—driverless cars, which seem inevitable—we need only explore the ramifications of a Humans and Autonomy Lab (HAL) insight, which bluntly states that, "The biggest challenge for driverless cars is not the technology, but the integration of the human driver into this deceivingly complex system."[1] Their point is simple and pragmatic—we are the challenge. HAL is mainly studying self-driving cars and those who have to try to function within them, the working assumption being there are no fully automated systems, however complex, and these automated systems, in order to function properly and efficiently, still require human awareness. In other words, we still need to have our hand on the controls and make sure the system is operating correctly. The system needs judgment. Automation will do just fine; the problem is how *we* are a threat to automation.

This is our brains on automated systems: decreased rationality, attention degradation, decreased awareness, drowsiness, complacency, reduced reactive time and loss of interest in one's surroundings. In short, the textbook definition of boredom. Assuming automated systems continue to proliferate in our personal, professional, political and emotional lives, and there is no reason to think otherwise, we will be increasingly tasked with monitoring these systems; therefore, the gravest threat to future humanity is not technology, but how to manage boredom and uselessness.

Numerous studies have drawn similar conclusions and it is a safety problem in some industries, such as commercial aircrafts, since lives are at stake with pilot performance. The solution, according to HAL, is to see if additional automated systems can detect states of boredom, or at least the incipient signs.[2] One can only imagine how this new boredom algorithm will work—it will monitor where the pilot is looking, their blood pressure, heartbeat, arm movement, agitation, durations of head swivels and drowsiness. If our patterns match the pattern of a bored person then, naturally, an alert system of sorts will try to enliven the person automatically (the "porn at work" phenomenon is related here, on a scientific level). The alerts will have to be sensate, of course, and not real stimulation; real stimulation goes beyond having cold air wash across one's face or being asked to do something or be introduced to color or having coffee offered, which are likely to be early solutions. But then, perhaps the solutions will become "true" stimulants—pictures of one's kids, riveting images or stories; after success in select areas, such solutions will be found in all workplaces.

In the scale of things the problem grows and the "threat" to the driverless car is merely an analogy. Once you *writ large* the problem cited here, and you apply the same effects of complacency and boredom to all of the automated moments of our lives, then you have a public health crisis. Everything is connected, as Muir said. Spousal abuse increases in a down economy. Drunk driving increases in an up economy.

The problem before us is how to shoehorn an individual into a *culture running itself* and how to manage the new threats to cultural production—us; or, more specifically, the fact that we get bored, disengaged and lose interest in our surroundings when all we have to do is monitor, thereby causing system malfunction; i.e., uselessness. We are the threat to culture, maybe not to how it is defined and performs today, but undoubtedly we will be a threat to how the future defines culture. We can be about as sure of this as water flows off a cliff. Boredom will be a public health threat, as devastating to an economy as the 2008 mortgage crisis or mass migration.

The Actual Numbers

Let us not cry wolf, however, and potentially overstate the problems of algorithms. What, exactly, are the researchers saying? To start, the 2016 World Economic forum meeting in Davos, Switzerland, has chosen automation to be its headliner and conference theme. An influential study from 2013, out of Oxford, concluded that 47 percent of US jobs, not just of the blue-collar variety, are at risk of being automated.[3] Jobs especially at risk are those in transport, sales and logistics. Jobs with a low risk of being automated include athletic trainers, dentists, clergy, editors, those involved in creative tasks, among a few others.[4] Those with social skills and involved in the hospitality or "caring" or nursing industries arguably have nothing to worry about, the assumption being we are just not ready for robotic nurses and the technology is not there yet. Though these stats were cited earlier, let's recap: around 40 percent of jobs in New Zealand are said to become irrelevant in the near future, 35 percent in the UK, 49 percent in Japan and about 50 percent of Indian IT jobs. In a 2016 study by Citi, coauthored by some of the same business economists as the 2013 Oxford study, more details were provided. For instance, we can learn that Thailand has a 72 percent risk of jobs being replaced by automation; China at 77 percent, Ethiopia at 85 percent, India at 69 percent. What is obvious is automation and its effects will not be distributed evenly, with some cities and nations bearing more of the brunt than others, mainly owing to the nature of the work in said region.

The economic effects can be seen in widening gaps of inequality and stagnate income for middle wage earners—often in manufacturing—thereby increasing concentrations of wealth in the hands of those who control the software. Some forecast a coming "algorithmic aristocracy." There are others, however, who claim the jobless future is a myth, a scare tactic, citing, for instance, the fact we still have bank tellers a decade after many predicted their demise in the face of ATMs (lower operating costs owing to ATM savings allowed the banks to open more stores).[5] For instance, between 1982 and 2012, the tech labor force increased significantly despite the increase of computers, outpacing the jobs lost to the computer. Clearly, the Industrial Revolution did not tank the labor force, but would go on to create jobs for some of the largest population explosions in history. Yet, automation does not have to erase jobs to create a "jobless future." Rather, it is the future of machine-directed tasks and monitoring of systems, which are jobs only in the technical sense of providing paychecks. Even if the automated future is one in which employment remains constant, the nature of the work will change (the substitution myth), already has actually, and this work will be largely in monitoring automated systems, with the exception of highly dexterous jobs and a few others. In short, the future of work in automated societies could likely be what anthropologist David Graeber calls *bullshit jobs*. Graeber writes:

> A recent report comparing employment in the US between 1910 and 2000 gives us a clear picture (and I note, one pretty much exactly echoed in the UK). Over the course of the last century, the number of workers employed as domestic servants, in industry, and in the farm sector has collapsed dramatically. At the same time, "professional, managerial, clerical, sales, and service workers" tripled, growing "from one-quarter to three-quarters of total employment." In other words, productive jobs have, just as predicted, been largely automated away…But rather than allowing a massive reduction of working hours to free the world's population to pursue their own projects, pleasures, visions, and ideas, we have seen the ballooning not even so much of the "service" sector as of the administrative sector, up to and including the creation of whole new industries like financial services or telemarketing, or the unprecedented expansion of sectors like corporate law, academic and health administration, human resources, and public relations.[6]

Graeber continues:

> Once, when contemplating the apparently endless growth of administrative responsibilities in British academic departments, I came up with one possible vision of hell. Hell is a collection of individuals who are spending the bulk of their time working on a task they don't like and are not

especially good at.

One could conclude bullshit jobs are not only going to increase, but also increasingly rely on automated systems, which will make them more bureaucratic and more boring, boredom being a state of "unfulfilled desire for satisfying activity."[7] Walter Benjamin, for his part, referred to boredom as "when we don't know what we are waiting for."[8]

So, as automation increases, we are going to get more bored, since we are not, as utopians claim, all going to fill our time with Shakespeare. Boredom is a moving target, like our moods and habits and preferences. As the latter changes so too does the quality and texture of boredom. But even though there are no proven tests to pinpoint boredom, we know when it lasts too long it morphs into sadness, then, if left unchecked, into depression or social disengagement. Drug abuse is a symptom of boredom. Boredom makes it difficult to identify your feelings. People who are bored become disengaged in their environments. People get bored when they lack control over their daily activities (blossoming automation). When you are bored, you blame your environment for your boredom, which goes hand in hand with the fact that boredom causes you to be disengaged from your environment.[9] When you lack the ability to entertain yourself, you are, or will be, bored very soon. Lastly, and nearly all researchers point out this phenomenon—people who are bored experience time differently.

One of the predominant causes of boredom is inadequate stimulation, which might strike one as odd given that we are on average exposed to 360 media ads per day, each engineered to pique your attention, ranging from text to video to image to graphic to what have you. However, when you wonder why we are trying to stimulate ourselves to such an *nth degree*, which we are with this great media experiment that bookended the twentieth century and continues to gather steam in the twenty-first, the answer presents itself. *Rather than being stimulated, we are under stimulated.* In a stimulation desert, we have created and consume systems of media exposure that can only be, on a conservative estimate, called sensory overload. It is a misaligned solution trying to solve the problem of under stimulation.

To be stimulated does not mean to be entertained, to have beautiful media put in front of you...that is to have your senses activated. On the contrary, to be stimulated means to be awoken, inspired, given air, to feel something deeply, perhaps without reason. Stimulation and great literature, or the early pangs of love, or the giggle of a child, or wild ravines in untrodden valleys, not only go hand in hand, but also mutually reinforce each other. We can be stimulated *through* our bodies and our flesh, but we should never equate sensation with stimulation. Stimulation does not have a place, it is not situated temporally, since to be truly stimulated is closer to being enthused, from the Latin *entheus*, possessed by a muse or god. Stimulation gives context while removing it, and it is not a simple bait and switch, but rather surgical, with the same complexities as a heart transplant. Stimulation gives qualities. This is why boredom, as the opposite of stimulation, is an "experience without qualities," according to Elizabeth Goodstein, and decidedly modern in the way we talk about it. Boredom is not so much a thing, but a way we describe existence. With the sudden appearance of the word *boredom* into the Oxford Dictionary, and its concomitant appearance in French and German literature, at its heart is an existential malaise, intertwined with alienation, ennui.[10] In some Christian theological traditions, boredom is considered a sin.

As to how automation specifically affects our workplace via boredom, this is a growing field of research, which also shoehorns into the discussion the "threat" some humans pose to automated systems; in other words, they lose focus and make mistakes because they are disengaged. For obvious reasons, one of those reasons being the future of humans traveling long distances in automated environments, NASA has authored its own studies about automated environments, boredom and complacency;[11] at this point, it is a vague problem with little in the way of solutions, but it is not a new phenomenon.

In a 1994 landmark report, the National Transportation Safety Board analyzed thirty-seven airplane accidents between the years 1978 and 1990, concluding in thirty-one of those cases — around 83 percent — the main point of failure was not "failure," but rather the crew inappropriately reading and monitoring the controls.[12] In other words, complacency, disengagement and the effects of boredom in automated environments. In 2016 a study involving the US Coast Guard found that, "With the increase in automation, boredom in the workplace will likely become a more prevalent issue for motivation and retention."[13] Across all industries and all jobs, boredom is a growing threat to the vitality and functionality of our labor force.

Yet, as it was with studying the emotional versus economic effects of automation, the latter more easily given to data and spreadsheets, how can one quantify the effects of boredom on how we relate to each other, view the stars, understand our bodies, fall in love or raise children — the stuff of culture? It is easier to study how boredom affects one's performance if they are a navy pilot managing automated systems, but to measure the effects of boredom on one's cultural performance — on families, how we engage in private or in public, with our children — is an entirely different matter.

Can Culture Fail?

We are still of value. Not useless yet. While human consciousness is still required to intervene and "run" automated systems, we will surely go the way of the telephone operator who once had to manually connect phone lines — now, of course, no such positions exist, as they have been automated away, and the statistics and forecasts project an analogous phenomenon across the globe, affecting different cities and regions with varying intensity. History has a way of defying prediction, but it is extremely likely that even if the jobless future does turn out to be a myth, the nature of the work is sure to change.

One of the most pressing problems in the future of automation is us, or, one of *our* biggest problems is learning how to work (and feel alive) in automated environments when the jobs become more rote, tedious, more monitor-ish and less doing. The techno-human crux is ensuring systems function flawlessly with flawed humans, since we, unlike our robotic kins, get bored, require stimulation, fall asleep, cannot focus when disengaged and get bored. Though it is just beginning to accumulate a critical literature, boredom is a serious affliction with destructive energies and boredom in the workplace is a threat to future productivity. If researchers know little about what causes boredom and what really goes on inside a bored mind, less is known about how to "cure" it. Currently, automated systems are being designed to detect human boredom in environments *already* automated and provide an automated solution. Is this not trying to put a fire out with gasoline? What happens when the proliferation of automated systems enters our personal life—and we become bored not just at work, but physiologically, intimately—boredom in your marriage or with your kids, but then provides a few solutions? In this sense, boredom is a serious public health threat, both to the cultures we create and to ourselves.

Can culture fail? What would be the metrics if culture fails? What does that look like? Can culture "crash" by user complacency like the thirty-one aircraft? One could argue, to the contrary, we are becoming *increasingly* engaged with the typical forms of cultural production—media—since individuals produce more content than any generation prior. Of course, it remains to be argued if digital participation is the same as cultural participation and one can only make this argument if digital culture is now the same as culture. Are we willing to say this?

In order for culture to fail, we have to establish a baseline of what culture is, since in research and academic circles it has the reputation of being one of the most slippery words in the English language. Without question, itemizing culture into subcategories does an extreme bit of violence to something so multifaceted and dynamic, but it does help narrow the focus, not to mention that there is no such thing as one culture, but rather the plural, *cultures*. Definitions can also get in the way—we do not need to define *what a child is* to claim that some can be spoiled and others can be well behaved. The same goes with culture. Nonetheless, painting in broad strokes, the thought experiment is something like this:

(i) Culture is *artificial* in the sense it is cultivated by humans or animals, and alive since the latter two are organic— inorganic matter does not have culture; while societies do not say, "hey, let's produce a culture," ways of being and *mores* happen spontaneously and creatively (with the exception of some top-down totalitarian regimes establishing culture);

(ii) Culture *serves* the living. Culture is embodied by social organization, customs, language, religion, arts/literature, forms of government, technologies and economic systems, among other umbrella terms attempting to describe what it is we do as humans, all of which help its members "thrive." Since culture is organic, an expression of the living, and since we are biological creatures who are hardwired to continue living and pass on our genes, simply "to let live" is not a satisfactory metric for culture; in other words, just because a *culture allows a people to stay alive* does not mean it is successful or worth passing on. We can grow an orchid in our bedroom, but this doesn't mean the orchid is living up to its full potential as a part of a vibrant, natural ecosystem.

Cultures must help people thrive in the *human sense of thriving* — we need to be stimulated to feel alive. Therefore, if one can make the case that culture is no longer cultivated by humans, does not serve the species, and does not allow its curators to thrive...then culture *can* fail.

Chapter 7

Poets for Hire

FBI agents who study serial killers create psychological profiles based on the nature of the crime, with the assumption that if you just connect the dots, a persona emerges, and once the persona emerges and you know what is motivating a killer, their behaviors and patterns become easier to predict.

At the heart of prediction is motivation—to predict is to know what motivates. Historically, human minds have performed these duties—priests, mothers, political advisors, seers, poets and election consultants have proven adept at delving deep into the human psyche and returning, often scathed, with a few pearls of wisdom. But is their time limited? Just as with Brad Pitt's character in *Moneyball*, is it not time we step aside and let computers do the work for us?

According to editor-in-chief of *Wired*, Chris Anderson, this is exactly what we should do. "Forget taxonomy, ontology, and psychology," says Anderson, "who knows why people do what they do? The point is they do it, and we can track and measure it with unprecedented fidelity. With enough data, the numbers speak for themselves."[1] Already, humanists are finding themselves within the world of software engineers since, as the myth goes, programmers suck at being human or, more kindly, they have often failed to cultivate, in their emotional lives, that bit of *je ne sais quoi* that renders us unmachinic.

This is why Silicon Valley is in need of poets. A *Washington Post* article sums it up nicely with "The Next Hot Job in Silicon Valley is for Poets."[2] Wait...*what*? Yes, poets and humanists are responsible for adding all those "umms" and "ahhs" and pauses to AI speech—Siri, Amazon, etc.—which, while completely unnecessary, renders them more quirky and relatable, with the desired end result being emotional connection and hence, brand loyalty.[3] The funny thing, it's working. The fact a select few poets can actually make money, however, is not an isolated case.

Algorithms need constant refinement, and one does not implement one and walk away. To the contrary, and despite their apparent effortlessness, their maintenance requires buckets of human labor, intellectual expertise and that human polish. Left to themselves, they can be quite stupid. For instance, when in a speech on the economy presidential hopeful Mitt Romney mentioned Big Bird, the phrase went viral, like Charlie Sheen's "winning," and when the phrase hit the Internet all of the algorithms treated it as a reference to *Sesame Street* and not politics. In this case, which is all too common, humans stepped in.[4]

A distinction needs to be made between a suggestion and a prediction. A prediction can be tested by its accuracy and lends itself to concrete phenomenon like voting patterns, physical movements and consumer choice, but a suggestion, as in a film recommendation or Pandora's playlist, has no surefire way to test its accuracy. A suggestion seems innocuous, but even it requires a predicative power to be a functioning product; if Pandora always played stuff that felt awkward and out of place, we would not use it. The algorithm, and hence the entire company, is successful because it has a high percentage of success. When the songs flow seamlessly, we do not notice and it has the appearance of good taste, as if curated for us, but it is really thousands of lines of code processing patterns and matches, except we do not see it. When it comes to something as malleable as taste, when it is not a precise solution, getting in vicinity of the target suffices, what we could call *soft consent*. Algorithms whose output are suggestions are soft predictions and their veracity cannot be tested, since in order to test the veracity you have to be able to replicate the conditions, and since you cannot go back in time, the efficacy of suggestive algorithms is based rather on flow—

does it generate a mediated experiential flow or doesn't it?

A specter of injustice hangs over all of these conversations like a giant cornice over a clean slope, threatening to crash. It is the specter of the ghost in the machine and of rationality itself. It is the debate about what makes us *us*, what defines us. Are we rational creatures who make irrational decisions, the latter mere mistakes in need of correction? Are we a mix? Are we mainly passionate animals who dabble in rationality? At stake in the proponents of the "we are a fully rational animal" is, in our current context, prediction. If we are fully rational and our consciousness follows logical structures (which we have yet to define) then, in fact, we *are* fully predictable; it is only a matter of processing speed and better neuroscience. If we are predictable, then we really are not creative, in the aleatory, happenstance sort of way.[5] If we are not creative, then we might as well outsource creativity to algorithms, since they are much cheaper. Sure, people will still paint and sculpt, but, if they find themselves awash in an industry where they are competing with bots, and consumers are unable to tell whose work is from a bot and whose is not, then, well, that changes the game.

The popular estimation is that we are an irrational animal, beholden to the whims of capricious, and often misguided, calculation. We overreact when we should not have, fall in love with strange things and strangers, and turn a blind eye when we should and equally when we should not. Drop a person in a new city and there is no way to predict where they will go, who they will talk to and what they will say in the first four hours. Like rolling a tennis ball off the top of a mountain, there is no way to tell where it is going to land. Right? Yes and no. New research relying on, of course, data analytics is beginning to paint a different picture and new concepts such as *mobility statistics, quantitative sociology, ground-state behaviors, non-mobility dynamics, human-life analysis,* can give an insight into where the research is being done on human prediction.

Ninety-Three Percent

Human behavior is 93 percent predictable, according to a group of network scientists.[6] Researcher Albert-László Barabási and his team first began tracking people's travel and movement habits in 2008, where they analyzed the movement patterns of 100,000 individuals with phones over a 6-month period (this is how they tracked them). They concluded that "After correcting for differences in travel distances and the inherent anisotropy of each trajectory, the individual travel patterns collapse into a single spatial probability distribution, indicating that, despite the diversity of their travel history, humans follow simple reproducible patterns."[7] According to Barabási and team, 93 percent of human movement is predictable.

Another paper, by researchers Alex Pentland and Andre Lui, took the study of human mobility to a smaller scale and ventured to see if they could predict subsequent actions from initial preparatory movements; in short, they were asking if people's movements can be predicted based on the first inklings of bodily movement. Pentland and Liu took driving as their case study, and discovered they could predict, with 95 percent accuracy, what a driver was going to do based on initial movements.[8] Why study this? The researchers want to help refine systems that can better predict our behavior, better anticipate our moves and, in general, better understand the connection between intentionality and mobility—which they considered to be equivalent to a "set of dynamic models." This research, like so many others, is narrow in scope but universal in application. If our future behavior can be predicted based on initial movements, then statistics can simply create algorithms to extend the temporal dimension from seconds, to minutes, then to hours. It would only be a matter of time before these formulas become constructive algorithms and thereby try to dictate human mobility.

Another group of researchers came to a similar conclusion following mobility trajectories based on smartphone tracking.

According to the report, "The patterns allow an enhanced predictability, at least up to a few hours into the future from the current location."[9] James Bagrow and Yu-Ru Lin, researching the social aspects of mobility, find themselves interested in how non-habitat patterns emerge (not your home or work) and how, akin to the spread of disease, there are implications for "the spread of information or rumors" across a social landscape.[10] Nathan Eagle, Alex Pentland and David Lazer, publishing their findings in the Proceedings of the National Academy of Sciences (PNAS), were able to predict friendship structures and even job satisfaction simply via tracking phone and call logs.[11] Having friends near your work, for instance, meant a person was more likely to be satisfied with their job, and frequently calling your friends at work indicated job dissatisfaction (they corroborated their results with self-reporting).[12]

Out of Stanford in 2014 came a paper titled, "Automatically Detected Nonverbal Behavior Predicts," in which members of the Department of Communication studied the relationship between creativity, problem-solving and nonverbal behavior.[13] What the researchers found is that rapport is a good metric to evaluate the potential success in an interaction. But the hitch is they are studying rapport solely on nonverbal, body posture movement and changes in the body in space. In other words, our body postures reveal our rapport with others and our rapport, in this case, can predict success in creative tasks; naturally, the researchers used algorithms as the engine behind their predictions. After their experiments, it was concluded that by using algorithms (among other things) to help analyze nonverbal behavior, the team was able to predict success in a "creative collaborative task," with about an 86 percent success rate.[14] Of course, once this technology becomes ready for game time, algorithms can be interfaced in creative teams to keep them on track and headed towards success.

We do not know yet what all these results mean, as scientists are simply exploring the problems they find interesting, and/or the ones aligned to research grants, but it is not hard to imagine how these results will trickle into culture: urban planning, disease control policy, social movement regulation, creative task-solving, advertising, behavior prediction. Early, crude algorithms will feel clunky, but they will learn and get better, and refine existing algorithms; new ones will be created and new predictions will be made, with new companies founded off this predictive power.

The Failure of Einstein

These are just a few examples of the state of human behavior predictability research. Historically speaking, the research is in its infancy — theories of gravitational force were first published in the seventeenth century and, 400 years later, we are still studying gravity and gaining insights about the movements of celestial bodies. If, during Newton's time, they could measure the size of planets, now we can determine the chemical composition of planetary surfaces hundreds of millions of light years away. If human predictability research has a similar arc, then the human will no longer be a mystery, and Yuval Harari will be right in his prediction that the manufacturing of bodies and minds will be the commodities trade of the future, since we will have mastered human gestation, gene editing and bodily curation.[15]

Human predictability research is also disparate, i.e., deployed in fields as disparate as advertising and neuroscience, criminal behavior and urban design, social media clicks and nuance in nervous speech. As more and more results and research methods become shared across specialists, and then into the halls of government and startup whiteboards, a race to the human theory of everything (TOE) will be underway; and yet, akin to Einstein's desire to unify the seemingly disparate laws of nature in a grand unified theory, the research will disintegrate — it will grow too many offshoots, too many specialists no longer able to speak the same language (like astrophysicists and quark specialists). This does not mean, however, that profound studies about human existence are not forthcoming, in addition to how those discoveries are applied. We only have to note that, *despite* the fact the universe is largely a mystery, we are still able to control a spacecraft with perfect accuracy, and predict all forces, millions of miles away in the blackness of space *without* a grand theory.

Full and exhaustive human predictability, however, is often embraced. Patrick Tucker, in his own musings on the future and technology, which he terms the naked present, argues that "the future is an improved present, safer, more convenient, better managed through the wonders of technology and invention"; "The devices and digital services that we allow into our lives will make noticeable *to us* how predictable we really are."[16]

But there is a lesson to be learned from the integration of supercomputers into the game of chess, which can now beat all but the best and most crafty chess masters handily—it was thought that the use of such immense computing power would lay bare the rules of chess and reveal hitherto unknown dynamics in the "greatest" game. Artificial intelligence via top-shelf algorithms was supposed to be a TOE moment for chess, but something happened, or did not happen. Artificial intelligence never upheld its part of the deal. Its moves and game strategies did not deliver, and while the machines can make great moves thanks to the ability to make a million calculations a second, it never revealed anything about *the game*. We risk the same type of non-revelation when AI is put to analyzing ourselves and we might be making more moves blindly in the future thanks to a form of intelligence we cannot understand.

Rara Avis in Terris Nigroque Simillima Cygno

Let us assume, as we should, that the computational power of software becomes unimaginable, and let us assume algorithms become more in tune to the intimacies of our bodies and minds, eventually in a decade or so becoming grafted to them with technologies we have yet to invent. Even if we assume that, indeed, all this comes to pass, we cannot say a human is predictable. When you are a hammer, everything looks like a nail. This aphorism about bias applies to reducing the human to decision-making—in other words, intelligence—and to the habit of computer scientists in thinking everything is computer-ish, or of researchers thinking that reducing the mind to a "set of models" captures anything substantial about our inner worlds. By the same token, we can accuse Walt Whitman of thinking the world is fundamentally poetic, or a mystic in claiming every part of our world is transcendent.

Harari is correct to note that driving a taxi does not require consciousness, but rather intelligence, and that automation, such as self-driving vehicles, needs only intelligence.[16] But this type of analysis, while correct from a task completion perspective (the cab ride), fails to notion the *quality* of the cab ride that is now automated and how cab rides are a form of consciousness; for example, every temporal point in human life is an opportunity to refine and engage our consciousness via other conscious beings, and if all we encounter are "automated intelligences" then the world becomes less conscious.

The impulse to interpret the world and everything in it through the prism of a single idea is "religious" in its origin and has consequences for the question of: can human thought be represented as an algorithm? In other words, it's the holy grail of cognitive computer science—making a computer indistinguishable from a brain or fully merging a brain and a computer. The *singularity*, as it was termed. To claim success in creating a computer that is really, actually, in point of fact a human brain, first you exhaustively have to know everything there is to know about a brain. Is that possible? Is the brain simply the sum of electrical firings? As of yet, we have no definitive answers, since we would have to wait until all the numbers are in from the divergent strands of research, and that situation is unlikely any time in the near future. More importantly, we need to take note of the *scale* of prediction research; in other words, despite the claims of human predictability, what cannot be said is that a human is predictable. What can be claimed is that within a controlled environment (a specific task, situation, etc.) there are statistical probabilities able to predict, with high levels of accuracy, what the course of action will be in said controlled environment (mobility studies).

On the other hand, there are strong cases to be made that human beings are just not cut out to be rational creatures. Formalized by Herbert Simon but debated in philosophical circles for millennia, *bounded rationality* is a theory of human behavior in economics claiming we, as finite and fallible human beings, do not always have the cognitive resources to make optimal solutions. The crux is processing information—we do not always perceive the right information and we do not have the capability to process all of the information we do have.[17] In essence, the claim asserts we do not make rational decisions and hence, our economic behavior cannot be predicted.

Abutting bounded rationality is the *black swan theory*, first presented by Nassim Nicholas Taleb, which claims that essentially unpredictable events in history—9/11, the Internet, WWI—have a disproportionate effect in culture; such events are unpredictable on account of the small probability of occurrence and because nothing in the past can reliably be used to predict their occurrence.[18] According to Taleb, so as to satiate our thirst for organizing chaos—a basic drive in human nature—we concoct theories for the black swan events *after the fact*; much in the way the rise of certain political figures in the 2016 US presidential election cycle seemed impossible to foresee until after their rise. What interests Taleb is not just this observation, but also protection against the black swan event—while what is a black swan event for one party might be a boon for another, the point is not to be surprised by the detrimental black swan event.

As it turns out, bounded rationality theory, if proved correct, is actually a good argument in favor of implanted AI algorithms — chips in our brains, etc. — in the sense that if it turns out we do display malfunctioning rationality (this is likely), then having the extra benefit of a super-processor to help us make decisions would benefit us. The same goes with the black swan: if AI can at least raise the statistical threat to human consciousness, such as of the rise of a dictator in a nuclear-armed country, then that is reason enough for having our brains enhanced by AI. However, the problem with AI integration is it rarely works this way — *after analysis* comes the power of prediction, then comes the power of intervention.

This trajectory has happened in so many technological industries it is hardly worth mentioning were it not for the fact so few see it coming. For instance, in education first came the data on student behavior online, when technology was ready to report on student clicks, login times, etc. Next, of course, as soon as the system became smart enough, companies started to create AI-based educational technologies able to mold to student behavior; in other words, the adaptive platform, which is the hottest thing in digital education. Analysis *to* prediction *to* intervention. There was the case of newsfeeds then, of course, the technology got so good that it could intervene on your feed and customize it. Automated devices for diabetes patients are already intervening on patients' blood sugar levels, whereas previously the technology could only report or analyze and predict likely future states. The analytical research into the structure of DNA began with analysis and, naturally, ended up with intervention and modification. It happened with agriculture (new bacterial resistant crops, for example), chemistry (plastics, gas refineries, for example) and the same trajectory with human behavior is inevitable.

Humans have always modified their environment through observation and study, and the rules of science are but a codification of that. Government and security ministers will soon be able to have sophisticated movement models and predictions for their most "dangerous" citizens, and preemptive policing will become irresistible when the forecast says that such and such an individual has a 97 percent chance, based on purchases, movement patterns, stimulation biometrics, etc., of attacking an elementary school. It is just common sense that when you understand the constituent elements of something, you take the next step in modifying it to your liking.

In fact, the trajectory beginning in data gathering to analysis and culminating in "intervention" also eerily maps to how humans learn and retain information. Learning scientists have long been studying the linear process by which we observe, read and retain meaning; then, after mastering the former, we are able to compare and evaluate; finally, we can analyze, synthesize and create (Bloom's 'taxonomy of learning'). It is a process whereby knowledge increases and our intelligence is exercised. Oddly, the arc of technological application is analogous in the sense that our early technologies grew from rudimentary fact-presentation functionalities to allowing for analytics and then "creation"; for example, intervention. What this means is not exactly clear, but since our intelligences are guiding software intelligences, we are becoming smarter with "smart," and the phase where we are able to create is when algorithms actively intervene in all facets of social and political life. That phase is coming soon.

Technological Agency

In the early stages of technology studies, say in the 1970s and 1980s when the Web was being formed, the question was asked if technology will ever have agency, the latter defined as *the capacity to act in a given environment*. Historically theorized mainly in the context of moral, theological and political philosophy, agency is closely related to willpower, free choice and ultimately consciousness, since something without agency is predetermined. Social psychologists, gender theorists, philosophers of technology, political scientists—everyone talks about agency. And for good reason: as a word attempting to wrangle with something so intimate as human freedom and self-determination, it is obvious why it is an essential term in the era of global democracies and independent nation states. It was only natural as our society began to become more technologically affluent and saturated, the question of *non-human agency* surfaced; in other words, technological agency.

Human thought and behavior, so the story goes, contains varying degrees of agency—just think of your free-thinking friends versus those who are more apt to step into the mold. Technologies presented early problems to thinking about human agency. Once we outsourced some of our capacities to early objects (wheels, rakes, horses), these non-human entities now had the ability to influence us; of course, peoples have been influenced from the *outside* for millennia, and the simple phenomenon of thunderstorms has had a profound impact on early rituals, religious formation and behavior, but in this case we are talking about potentially self-generative, self-forming cultural objects able to overwhelm a social landscape given the right economic and political conditions (those "conditions" define us today and this book has been attempting to trace what those conditions could be).

A handful of excellent counterarguments exist and an oft-repeated one is that as machines get more powerful, so do we *by default*—the "intelligence amplification" argument.[19] We will not be replaced, but brought up with the tide. This is a good argument (*trickle-down AI*) and seems to hold true for computational procedures or analogous tasks—yes, having a brain-computer interface could help some drivers who cannot see, aid some scientists to develop better cures and already we are using "soft" AI when we use navigational apps. The advent of the camera, however, did not just allow us to store pictures other than in our brains, thereby freeing up precious "space," but altered habits of memory, and not always for the better. Assuredly, intelligence amplification is true insofar as a certain type of computational intelligence is amplified when minds work alongside software. Does intelligence become *more intelligent* by virtue of becoming more powerful? This assumes, of course, it is the destiny of "intelligence" to be amplified; in other words, it is what intelligence wants. There is no argument to support that claim.

A case can't be made either way, yet it seems fair to say intelligence reaps a reward of sorts when linked to software, disregarding the *atrophy affect* (losing an ability because you rely on computers to do it for you). Further, we must not forget that intelligence, as part of consciousness, is tied to a body. Can we say our bodies feel more joy—the baseline here—when it is armed with more intelligence? This claim likewise seems hard to swallow. Still, for argument's sake, let us concede intelligence benefits. Is there a parallel culture amplification? Does the vibrancy of culture(s) receive a net benefit? One thing we can no longer assume is that the application of AI will be limited to decision-making procedures, since, as of 2016, the genie is already out of the bottle and into our (biometrically smart) lingerie. While intelligence might benefit in the short term, it seems the case everything else loses in the long, simply because the long term traditionally relied on human intelligence. Technology working actively and independently in culture is *technological agency*.

What, then, defines technological agency? Nothing, really, insofar as *no one* thing defines human agency. We are a fickle animal, animated by strange phenomenon at strange times for unpredictable reasons. We fall in love, we hate, we trundle into depression. A meme floating around in cyberspace, like the goose that nearly downed an airliner by flying into the engine, can take multiple paths simultaneously, affecting people, other algorithms and real or virtual events in ways no algorithm can predict; or, if an algorithm could predict it, then scientists have about zero confidence they could build something like it.

Life is and has always been messy. A better question needs to be asked. The question is not what defines technological agency, but what *type of environment* is created by technological agency. This is the question of the texture of existence, of culture, our shared life and how our joys are formed—whether we like it or not—in bodies with a sub-100-year span (the latter range to be continued, of course). While many qualities of environment surface, two in particular are relevant—*complexity* and *controlled*—as essential to thinking about how the entire panoply of algorithmic procedures affects the vibrancy of cultural life. As to the question of what makes a culture vibrant, we need to sidestep intellectual definitions and rest assured, you know it when you see it. It is a felt phenomenon, inspiring, illuminating and, most importantly, stimulating (but not of the cheap sensate variety). While a *controlled* environment and a *complex* one appear to be on opposite sides of the spectrum, they are in fact mutually reinforcing.

We live in a world where we are surrounded by complexity. Data travels in ways, packages and methods less than 1 percent of us understand. Writer Sam Arbesman asks a prescient question when he wonders if we have finally created a world that is indecipherable to us.[20] Well, of course we have. Arbesman cites the complexity of modern civilizations and their infrastructure as examples—300,000 intersections in the US with traffic signals, to which no human mind can comprehend all the trajectories, patterns, risks, etc., much less operate all the controls even if they did understand it.

A familiar forgotten password example:

Enter the name of your childhood best friend? Hmmm. Would I have spelled it Phil or Phillip? Damn. Which one was it? I only had one attempt left. Ok, here goes: P-h-i-l. Yess!! What was the name of your first pet? Shit, I never had a pet. I couldn't have chosen this question. I did have a fish, and a hamster. But would I have considered them a pet? Which came first? Whatever. I enter D-o-l-l-y, my first pet fish. Success again. I'm in—password reset. Shit, my credit card has expired. That part is easy. On my way now. Shit, my device isn't being recognized. Too many people on my wi-fi? I restart. Do everything over again. Goddammit—I must have given my logins to too many people. I need to reset my password again. I'll do it on the road.

Simply explaining how your Apple ID *really* works is a mystery, much less how texts travel to space and into satellites, and then into other people's phones is basically inexplicable, unless you are an expert, and I doubt one single person understands each step. All around us, with our tools and processes, the world is indecipherable. As data grows, so does the need for computing power and as the latter increases, more data can be gleaned, which means we need more computing power. It is a veritable positive feedback loop. The Panama Papers, leaked in April of 2016, contained around 11 million documents—it would take a small government over a year to sift, organize and log all of that data (not including the interpretative work); yet, with a few custom search parameters and some acute data-modeling techniques, we can now search with ease, just as WikiLeaks does with the Clinton emails or NSA Web spying. Data keeps growing, but the ultimate platform upon which it functions is limited in scope, hence one reason why automation exists is because it *makes sense* of data. As regards outgrowing one's native home, automation is now making sense of things where we did not know sense existed (the Pop-Tarts and hurricanes example), and it is like a giant celestial net capturing the sparks of human industriousness (and boredom), analyzing their activity, then, of late, *modifying* that activity in the form of artificial intelligence.

The US legal code has 22 million words,[20] which comes to about 73,000 pages; not to mention the fact that when my antivirus finished with my computer it said it scanned 354,788 files. I use only about 20 files per day, maybe 30 on a Wednesday. Yes, the rest are necessary, but I do not know how they support the stuff I do, only *that they do*. In fact, I understand very little about how our world functions. I do not know how my alarm clock works. I mean, I know how an hourglass works, but how is time kept digitally? If all of the world vanished and it were up to me to recreate them — analogous to the way the monastics protected knowledge in the Middle Ages — alarm clocks would never be reinvented. Same goes for my microwave, my car, my computer, the chemical compounds in my water bottle.

The world is exceedingly complex, more complex and anti-intuitive than any other time in history; this does not mean we are more ignorant, but it does have epistemological consequences in terms of expectations — we do not expect to understand how things work, only that they do. We do not *expect to understand* the basics of building engineering, only that we are safe when we sleep and the roof does not collapse. However, "that they do" is no form of qualitative judgment and a cop out. This type of epistemic non-knowing, resulting from complexity, is in part what is fueling our indifference to the proliferation of algorithms in our personal and emotional lives.

Computer scientist Danny Hillis has claimed that we "are at the dawn of the Age of Entanglement."[21] A clever moniker, and twist, on the well-known Age of Enlightenment, according to Hillis the Age of Entanglement is actually the former's final expression. The Age of Enlightenment is a catch-all phrase for European developments in science, natural philosophy and generally ways of governing from the early seventeenth century through the close of the eighteenth century. In its essence, it put the quest for rationality, and human reason, at the center of human inquiry, and this rationality was sought in the cosmos, human affairs and government; if rationality and order were found to be lacking in said areas, it was desired to apply it accordingly. For obvious reasons, we remain wedded today to the motivations of the Enlightenment era, beholden to the processes of logical control in government, science and the cosmos.

While our world is impossibly complex, we allow it to be so because we think, presume actually, it is under control. The logical structure to the universe, prescribed by the scientific method, gave the world the impression there is an autonomous layer of rationality outside *in the cosmos*, independent of our minds. According to Hillis, in turn, then, computers, as the "ultimate expression of logical deterministic control," allowed us to create a further complexity and autonomy beyond our own minds, one that does not just exist independently, but can operate and intervene independently.

According to Hillis, however, with the properties of *emergent behaviors*, thanks to algorithms, the first crack in the Enlightenment foundation was witnessed; emergent behaviors are behaviors that were not part of the properties of a system's simple components. *Ex Machina* (2015) illustrates this perfectly when the android starts to display behavior not intended in the original programming. Emergent behaviors are more complex, and hence more unpredictable, than the sum of their simple parts. Other cracks ensued.

Hillis does not necessarily worry about the entanglement, however, since the new forms of human-machine collaboration are often rich, unpredictable and sometimes superior. Hillis has a point—the idea of humanist (and enlightenment) creation putting the human mind at the center of life and creative expression is an idea not reflecting what is going on today. Yet, we should not spill the milk while reaching for the cereal; in other words, the flip side of entanglement is uncertainty and should entanglement ooze outside of exciting new objects, such as mash-up code or experimental music, and into our personal lives, you can bet that the entanglement will go from beloved king to hated tyrant when we cannot understand what is going on in our bodies and minds because we cannot understand what is causing the interference.

Complex environments produce a type of *anti-knowledge*, like the omnipresent anti-matter of the universe that we can't detect, but the former is not always negative, just in this instance, since one *can* be in awe of complex environments. Ecosystems are complex. The night sky is complex. Complex environments are nice in the abstract, as semi-aesthetic objects of reflection, but when you are on the nether side of complexity the physiological effect is far from awe.

Complexity is a result of the entropy of civilization (a debatable point)—a product of nation state management, populations, mass sanitation, etc.—and software is the solution to complexity management. However, complexity and the perception thereof has moved from the "back end" to the "front end," from behind the scenes management to where it is experienced personally every day.

While we are not conscious of complexity during our day-to-day interactions, it is still consumed—even more so—in our physiology, in the metaphorical manner of when you are walking through a dense, prickly forest you are not conscious of each thorn on your flesh; nonetheless, you bleed just the same. It is therefore becoming increasingly visible and hence, more liable to affect our behavior, which, in turn, affects how we compose the world around us. This *would* be a tolerable situation were it not for how humans respond to complexity. In the face of complex environments, we stall. We are unsure what to do. We become disengaged. Less moved. Concepts of ownership are altered in complex environments and we do not see ourselves reflected in them. Complexity, while having an infinite number of entry points for empathic connection, surprisingly has few. Complex environments feel closed, even fragile in their robustness. Our understanding becomes frustrated. Complex systems feel autonomous, degenerate even, in the sense of being the product of maladaptation.

Simple animals, to the contrary, are some of the most complex entities on the planet and we are aware of that, but they hide their complexity well, since complexity (and hence, the entanglement) is equally about perception, which has to be a reason why in the face of some of the most complex technology ever constructed, Steve Jobs insisted on a Zen-like interface to Apple's products, to put the user at ease, to make them forget the frenzy of data traveling beneath their fingertips.

In complex environments, we do not know if we can contribute, because such environments feel inhuman in the sense of being overbuilt, overthought or under thought...we feel things are out of control because we cannot control complex environments in any simplistic manner; we become disengaged, bored. In politically complex environments, we feel out of control, angry, and want to revolt; this translates to irrational elections and selections of leaders. When we become bored, we need to be stimulated, hence the spectacle of media societies. Since we are unengaged, new forms of intervention are invented to make up for a lack of contribution; in other words, automated creativity.

Complexity breeds new forms of management, and algorithms enable this management. More than just new forms of management are created, but also new types of environments; just as when large numbers of migrants settle into new places, the culture of that place changes on a material level, what is eaten, how people greet each other, how relationships are formed and so on. When taken to the level of culture—the nebulous sum total of our creations and creative processes—our environments become akin to *controlled environments*; a controlled environment, a phrase borrowed from research methods, is to be understood as the ideal place where hypotheses can be tested, as in a lab experiment. A controlled environment is valued because it is controlled, so results can be repeated and thereby verified. If an environment is not controlled, results cannot be verified since a variable or variables could have entered the experiment and altered the outcomes.

Algorithms, as social experiments, work best in a controlled environment. If it is increasingly the case software is determining the character and construction of our world around us—from our media to mate selection to smart cities—which it is, then it is only logical that it will "seek" out ways in which to become more effective. In other words, our world will become a controlled environment, which has ramifications for every part of culture we experience, as well as experience itself.

What do algorithms seek? To become more effective. An algorithm lives or dies by its effectiveness. If Pandora played music completely unaligned to your taste, or the last song, the company collapses within a week. The terms *positive* and *negative* feedback loops are operative here. Positive feedback loops occur in a system when an output becomes a new input and a change is amplified as a result. For example, when a stock portfolio gains interest, the account increases in value and when the account increases in value, more interest can henceforth be had, and so on. This is a positive feedback loop, increasing productivity. A negative feedback loop brings forth a decrease in function, often just maintaining system stability. When your body gets cold, it shivers to increase heat in the body and therefore prevent a further decrease in body temperature. Negative feedback trends toward system stabilization, while positive feedback loops trend toward amplification.

The interface between technologies and culture assumes a variety of forms and *we* can easily act as constraints in a positive feedback loop when we decide, as a people, that fully automated matchmaking needs to be stopped. In the case of culture, positive feedback, while aligned with growth, is actually a decrease in opportunities, which can also be understood as a decrease in variables, hence closer to a controlled environment. This is because technological agency produces limited fields of action.[22] If software continues to construct a cultural landscape wherein their logic and products dominate—which is the case—then by default a controlled environment is constructed, since the net effect of being offered *less options* in general and more options from the *software itself* creates an optimum environment for the exercise of the algorithm. Like a Trojan horse contractor, it builds and creates amidst us.

What is remarkable is how the intervention of smart machines into the milieu of culture has the opposite effect of what intelligence is supposed to do. What intelligence is supposed to do—what it affords those who use it—at least according to an influential theory by Alex Wissner-Gross, is increase *fields of action*.[23] According to Wissner-Gross, intelligence naturally fixates on future potential and, in many ways, his theory of human intelligence reflects the techno-optimism shared by many in the data sciences. He writes:

After all, technology revolutions have always increased human freedom along some physical dimension. The Agricultural Revolution, with its domestication of crops, provided our hunter-gatherer ancestors with the freedom to spatially distribute their populations in new ways and with higher densities. The Industrial Revolutions yielded new engines of motion, enabling humanity to access new levels of speed and strength. Now, an artificial intelligence revolution promises to yield machines that will be capable of computing all the remaining ways that our freedom of action can be increased within the boundaries of physical law.[24]

Wissner-Gross may be correct regarding the application of technology to tools and processes, but this is to miss the point—the application of intelligence clearly limits fields of action in social, creative and personal areas, and hence, what is traditionally called human "freedom"; human freedom has seldom historically been comparable with the ability to alter one's environment, but rather concerns moral and ethical fields of action. Our cultural field of action will be limited in terms of what is consumed (algorithmic options), what can be produced (the limits of automated creativity) and what reactions can occur as a result of both (what benefits, or lack thereof, we feel from our productions and consummations).

Shards of Frustration

A bestial and indefinable affliction.
– Dostoevsky on boredom

Research might soon prove how predictable humans really are. Large data sets will accumulate in the sciences and in the social sciences, and, like theories of gravity, a nuanced theory of the human will be pieced together; while there will never be a TOE of the human, there need not be one to make accurate predictions, just as you do not need to know the nature of the cosmos to land a craft on Mars. Already, scientists can predict human mobility with extremely high degrees of accuracy (93 percent), and the ability to predict how we think, feel and behave is gathering steam as one of the last great scientific questions. Not only a scientific question, but also one of marketing and entertainment.

The research community is just past the data-gathering era, in the middle of the analysis phase, and beginning to enter the intervention phase; the latter is when, armed with prediction research, culture is a place where software intervenes on us *and* our choices, and we and culture are altered as a result. As the intervention stage is underway, our cultures will become more complex, indecipherable and incoherent to us; the emotional consequences of this trio is disengagement, lack of awareness, boredom and ultimately feelings of uselessness. When a population is disengaged, solutions are drawn up to keep culture going, hence why we are witnessing the birth of an automated culture (technological agency) and political irrationality simultaneously; it is not a conspiracy, just a pragmatic fact of trends in business, technology and our own psychology coming together. Ironically, more intelligent ("smart") software in the cultural sphere will give us less options and less humanistic options to boot, effectively curtailing human freedom and making us less intelligent; we are not becoming less intelligent because we will think less, which will undoubtedly be the case, but because our fields of action will become narrower. Given that human consciousness is a balance between emotional and rational processes as we confront situations in time—intelligence is thereby constantly growing, in motion—we become less intelligent, and less free, by virtue of having a myopic field of action; a concrete example is the type of labor we will be performing in the future. If there is a formula, it is: what we gain in technical intelligence we lose in social and emotional intelligence.

Culture will become lifeless. Culture will become machinic. We will become like robots—such fearmongering is pointless. Culture is not a thing out there we can identify, but a multidirectional feedback loop constantly in motion. Everything we do, eat, talk about, love or hate constitutes and informs culture. What we are talking about with the word culture, as controversial as it is, are situations in which we live. We are talking about moments with emotions attached to those moments, since emotions live *in* time and are *of* time. To imagine the future is a literary enterprise, but this does not make it less true. The future will be us living on, just as we are today; that much is for certain.

What we have to decide is if how we configure our future is tolerable. That moment in the future where parents hug their kids less, because biotechnological fabrics connected to software detects their needs, then satisfies that need instantaneously. That scene when two teenagers are put together by a program and believe, in the deepest part of their psyches, that this is the person for them; only to be startled, weeks later, when a new match appears and they leave each other. We will be racehorses put to work as mules, built for something else, and our stock and trade experiences will be built off dissatisfaction, projection, frustrated libidos and thirsts for whom the watering holes have all but desiccated.

Everywhere, like malware on a computer, shards of frustration will transform themselves into aggression in the global economy, and it is already happening now in the form of terrorism and political irrationality fueled by feelings of uselessness. The irrationality of choice is only going to increase. Yes, we will become more placid, satiated, and a new era of "peace" could descend, but behind all great peace lives even more profound cruelty. And think again if you think of boredom as a harmless daydream. Uselessness, and its cousin boredom, is a fierce slingshot of the worst of human energies, the Janus face of desperation.

Part III

Nature

Chapter 8

The SMART Planet

On your body something grows without restraint, though you restrain it. It could grow forever if left to itself. There is even a fairy tale dedicated to this thing. What is it? It is your hair, willing and able, unlike your skeleton, to grow continuously throughout your life.

Given the right conditions, some animals are capable of what biologists call *indeterminate growth*. A large number of fish and snakes are indeterminate growers, and some parts of our own bodies exhibit the same patterns. Indeterminate growth means that something keeps growing and never stops, sometimes to its own demise. It does not mean it will simply consume everything in its sight like some sort of ravenous monster, only that the physiological regulations are somehow "turned off" in these organisms. Such organisms can grow and grow and grow. The human skeleton, for example, stops growing at a certain point, yet that of a kangaroo keeps growing until the day it dies. In some instances, what enables this unrestricted growth is their environment. The medium of water, for instance, does not put the same strain on a skeleton as it would for land creatures and so, growth has fewer constraints for aquatic life; the few human cases of extreme growth exhibit the fact that, as it stands now, our tendons and ligaments cannot tolerate the strain human activity requires, which includes the constant fight with gravity. Some plants, such as tomatoes, can be indeterminate growers.

Is human consciousness an indeterminate grower? But how can we make the leap from a biological specimen to something as amorphous as consciousness? Do the same rules apply? No, the same rules do not apply. It is merely a metaphor, but it is a fertile one. Also, the brain does not really "grow" in the manner of an organic indeterminate grower, but applies its faculties so that its *power* grows; *power* is to be understood as exerting influence. What, then, is this type of power (no, it is not "rationality") and what does it want? And what resides in our brain's constitution, or current culture, that makes consciousness a candidate for indeterminate growth?

Let us be clear, the brain is not an organism, but it is an organ, and while organs work with other organs for the betterment of an organism, our brain has exhibited the most profound, even reckless, traits of indeterminate growth this Earth has ever witnessed. One could say the indeterminate growth of our brain's power — automation in the form of "thinking" devices — is working in direct competition with our own power. It is not an irrelevant observation to make that businesses outsource to scale their operations; for instance, if you want a project requiring twenty-five programmers for 6 months, then it makes sense to obtain those resources outside your company, since you do not have the labor, cannot afford to make new full-time hires and also, the project is over in 6 months anyway (the rationale for the infamous "gig" economy). Outsourcing can also be cheaper than doing the work in-house, such as sending data-entry projects to India because their tech teams cost less. In both instances — scaling and cost — you can also witness growth *and* a decrease in in-house technical knowledge; namely, regarding the latter, the organization outsourcing loses the opportunity to develop in-house technical knowledge. This also explains the growth of the job title "project manager" who, while not always knowledgeable of their fields, are able to resource those who do. So, it is only natural that one of the biggest trends in business operations is to be small, nimble and agile, and yet "do everything." In this case, which is slightly different from human intelligence, the business suffers not and this is where the analogy breaks down. While it is not detrimental to a businesses' collective intelligence to outsource, it is so for a human mind.

A counterargument is that AI will never really grow to such a powerful, intelligent decision-making force in our world, because we have no working definition of intelligence. Kevin Kelly, a founding executive editor of *Wired* magazine, argues this precise point, while admitting that processors and sensors are increasing exponentially, stating that we need not worry about "exponential increase in the output intelligence because in part, there is no metric for intelligence."[1] Kelly may be correct in that as regards fully supplanting our intelligence on an operational level, it may be a pipe dream, but he fails to distinguish its "soft" intelligent outputs, such as in realm of culture. Also, more bluntly, it seems hard to argue that AI has not grown exponentially, since the developments in tech simply paint a different picture.

Hunting Traps

Aspen trees and grass—these are indeterminate growers. Indeterminate growers have an innate physiological ability to grow and their environment enables this. So, what about consciousness thought makes it able to grow indeterminately? It is able in the sense thought is ethereal, existing nowhere but in a body, yet divorced from it somehow, in some manner. The nature of thought—*what is it? Does it exist independently?*—is arguably the oldest philosophical question, beginning with the pre-Socratics formally, but also found, masked and mythologized, in the art, stories and culture of earliest humanity.

Consciousness has an ability for indeterminate growth because nothing, it seems, can limit thought; at least we have not found reasons to claim otherwise. It is a strange medium, or thing, or electricity, or God-given trait, or force, or power—we still do not know what it is. Verily, throughout history, it has proved extremely malleable, an agent of peace and equally an agent of murderous calculation. It is the last great frontier and it is no coincidence scientists can now land machines on comets rocketing through space, as they did with the *Rosetta* spacecraft in 2016, and make minutely accurate predictions about gravitational torsions and cosmic forces, all the while accurately predicting how your teenage daughter will cope with her boyfriend's infidelity will likely forever elude us.

Consciousness has always had *ability*, but it never really had the *environment*. Historically, in order for us to deploy intelligence, we had to be there. Rarely could we leave intelligence behind to act on our behalf, after our bodily presence had left. Artworks are a fragment of human intelligence we divorce from ourselves and send into the world, and primitive hunting traps are another form of divorced intelligence able to act, with intelligent intent, after we have constructed them and left them be. Yet, the scope, scale and power of these forms of *circulating intelligences* pales in comparison with what is happening today; in other words, the right *environment*, along with independent technological agency, is a recent phenomenon. Ubiquitous computing—which is integrating software into all facets of not just culture, but nature as well—is the desire to externalize and "grant" our intelligence to things that we believe—since we are handing it over—do not have it.

Ubiquitous computing is practically expressed in IOT or other neologisms, and fueled by automated algorithmic procedures. As regards granting intelligence to everything, one place where we can find this externalizing impulse on great display is in the idea of a "smart Earth." Just as companies mine social media data—what we click, search, like or buy—companies are seeking to mine *nature's data*. Big data supercomputers are modern-day open-pit haulers and rather than extract lime for concrete, they extract data for…for *what exactly* is the confusing part. It is in this sense that data, according to IBM, is "the new natural resource."[2] But, what does it mean for the Earth to need our intelligence? And what does it say about a culture who thinks the Earth needs *its* intelligence?

The Smart Planet

IBM was ahead of the curve in 2008 when they launched their Smarter Earth campaign (recently rebranded in 2015 as Cognitive Business). The goal was to use big data analytics to analyze efficiencies at the highest level of global industries and then, with that data, make strategic interventions. In an interview, the campaign was described by Rich Lechner, at the time IBM's Vice President of Energy and Environment, as such:

> You can look at our distribution systems around the world and see that more than 20 percent of all the shipping containers and more than 25 percent of the trucks moving around on a global basis are empty. You look at the way that food is distributed and understand that the average carrot in the United States—the lowly carrot—has traveled 1,600 miles to get to your dinner table, and you say clearly something could be done to improve the efficiency of our food distribution system.[3]

IBM's Smarter Earth was, after all, not quite about the Earth but about finding inefficiencies in infrastructure development. After IBM's initiative, Cisco followed suit, as did Google, Siemens, Ericsson, ABB, Apple and virtually every other big tech player. Then, in 2009, HP announced CeNSE—in other words, a self-proclaimed "Central Nervous System for the Earth"—which took the smarter planet from cultural and industrial ecosystems to actual ecosystems. HP proposed a vision of billions of sensors in everything imaginable—trees, water, streams, soil, mountains, etc.—retrieving the data, analyzing it and using it to make automated decisions about everything from ecology to rain to pollution to climate and so on; again, we see the same learning-action model here of data collection, analysis and intervention. It sounds great on paper, with HP describing it in the following manner:

CeNSE consists of a highly intelligent network of billions of nanoscale sensors designed to feel, taste, smell, see, and hear what is going on in the world. When fully deployed, these sensors will quickly gather data and transmit it to powerful computing engines, which will analyze and act upon the information in real time using a new breed of business applications and web services.[4]

Ironically, or not so ironically, HP's first major customer for this initiative was Shell, who will be using the earth CPU for extracting oil. Ideally, according to HP, the sensors would mimic the five human senses and be able to stream its "sensations" to remote clouds, which could in turn provide real-time data about oil behavior, or lack thereof, under the soil. For many industries and corporations, a global earth IOT is the low-cost solution for managing traditionally high-cost decisions, such as oil exploration. Or, for instance, is this year a good time to increase the oyster harvest in Maryland's Chesapeake Bay? With thousands of sensors embedded into the water and oyster reefs, and the ability to analyze said data along with historical growth patterns, water temperature, water quality and a host of other data points, a smart Earth algorithm could easily tell you when is an opportune time to harvest oysters, and how many, where and what size. Of course, a seasoned oysterman, or oysterwoman, could do just the same, but they would require benefits and a salary, and quite frankly, cannot be everywhere at once like the sensors.

And yet, smart earth initiatives may be our last defense against what appears conclusive: a changing climate in which sustaining life is more difficult. Remote sensors are tracking receding glaciers, water quality, frequency of storms, water levels and endangered animals; this data is a weapon in the hands of the scientist since, once it is modeled, assembled and conclusions drawn, it can be taken to the public. Seemingly but a continuation of animal collaring, the marriage of data and ecology are helping understand the migration patterns of wolves to aid in their reintroduction, or save endangered species, not to mention endangered ecosystems. Increasingly, it is claimed that big data tracking, once embedded into ecosystems, will be our Earth's *only* savior. With the managed protection of animals, food drops, tagging, caging, rerelease and so on, software is already being used to manage our ecosystems.

The future will bring us eco-notifications from as many species as we can track. Did a flock of birds that usually settle in Japan for the summer just end up in North Korea? Yes, the data tells us so — we just got pinged. Immediately, correlations in the air temperature and pollution and sound conditions in Japan could be the result. Well, the new birds now nesting in North Korea are, via their droppings, bringing non-indigenous species of pine to the mountains. This new species of pine are water hogs and light blockers, and will, so the data says, kill the native bushes the deer feed off. However, our data tell us if we can remove the sound pollution in Japan and make the environment more pleasing for the birds, and also spray a chemical in North Korea the birds do not like, then maybe they will return home. With the data and fearing the predicted loss of certain animals, the North Korean government sprays the forest.

What is essentially a cause and effect analysis, this can go on forever. But is big data up to the task? Some biologists fear we are falling victim to the Lego fallacy, which is "the idea that ecosystems are like those Lego sets where you build the Millennium Falcon and if a piece goes missing all you have to do is find a replacement and pop it back in."[5] Regardless, just as there is no such thing as raw data, there is no such thing as *parts* in natural systems and just as we do violence to our sensate world when we turn a road into pixels so software can interpret it, when we compartmentalize nature's parts owing to a systems management approach, nature loses the ability it once had to act holistically; i.e., its intelligence.

On an Earth where everything is connected and where the butterfly effect seems to be evermore the case, remote biometrical devices are surrogate human eyes and ears, able to predict and in some instances prevent major catastrophes. On up AI goes— from smart bedrooms (lights, alarm clocks, etc.) to smart houses (thermostats) to smart offices to smart cities to smart parks to smart countries to a smart Earth.

Internet + Internet of Things = Wisdom of the Earth

In 2009, then Chinese Premier Wen Jiabao gave a series of public speeches extolling the virtues of IOT, advocating for its rapid development and promising funds for government research. In a cryptic formula, he provided a distillation of the industry perspective of how IOT can help nature do what nature does best: "Internet + Internet of Things = Wisdom of the Earth."[6] Whose wisdom? What is this duplication of wisdom, since the Earth already has a wisdom of its own and can, presumably, get along just fine by itself?

What is problematic is not so much the existence of terromatic sensors, but how our perspective of nature will change as a result, much as we will approach culture differently if we thought it was simply the product of automation. Perspective matters: perspective is a *revealing of truth*, a confession or a violence done to the thing. All it takes is the person sitting across from you at a restaurant to tell you that the waiter dropped your steak on the floor—"I saw it when the kitchen door swung open"—for you to lose your appetite. The proverbial steak on the floor, in this particular instance, is that natural systems *need our intelligence*. When humanity is positioned as managers of a nature needing it and not as visitors allowed to visit, then nature understood as a vibrant, living, independent phenomenon falls by the wayside. It may be the case there was never such a thing as true wilderness, it being more the result of a romantic yearning, since our lands have always been populated, but mass perspective matters.

Verily, each animal, plant and stone is already a living sensor designed from millions of years of life to transmit, with maximum efficiency, vital chemical and biological components. The Earth is intelligent, without our contribution. We do not need an intelligently designed Earth for the Earth to be intelligent and if we think so, we have in part found one of the sources of our pathological scramble to give the Earth an intelligence—in other words, *we do not think it has any*. 'Tis but an instance of indeterminate growth.

So, if a smart Earth is the inevitable application of AI, new questions arise—who owns the Earth's data? Can the Earth's data fall under copyright, just as some genetically modified organisms are? Privacy is an obvious question and in the era of big data it is no surprise the hottest commodity in the future may not be data, but privacy. Already, keeping Google Earth out of neighborhoods is a luxury the wealthy are paying for. And to only complicate matters, in 2016 New Zealand made it possible, in legal terms, for a body of water to be considered a person,[7] which brings up the question—does a body of water, which is a person, own its data the way a person might? What legal recourse can be had to obtain it? What happens when it is violated and who sets the terms for defining what a violation is?

The idea of the Earth having an intelligence, however, is nothing new. Philosophers, since the inception of the discipline, have claimed a working intelligence to the cosmos, a *logos* (loosely translated as rationality). The *logos* was a fluid concept in antiquity, varying from thinker to thinker, but the general consensus had it human intelligence was but a slice, or reflection, or spoke, in the more expansive *logos*. The second-century Roman Emperor Marcus Aurelius believed in the *logos* of the world. The duty of human rationality was to acutely discern the rationality of the *logos*, so we could live better. Taoism has its own brand of a mystical rationality governing the cosmos and humanity, and Hinduism would personify divine energies as Brahman. In short, civilizations have long thought the Earth had a wisdom, often for our benefit, but it was always us who had to listen to it directly, through our bodies, emotions and history; the Earth's wisdom was never operational, never programmable, but always human in a way.

In a fitting move by algorithm writers desirous to capture the intelligence of nature, many in the computational and data science industries are actually looking to nature as primary source code, wherein computer scientists are analyzing nature for *its* patterns, *then* translating those patterns into code, then, finally, applying the algorithms to wherever they are useful.[7] But, again, this strategy is not new in kind, only in application. The Golden Mean of antiquity was an early attempt to translate patterns in nature to patterns in human life, often to achieve a desired end, such as harmony. Buildings across Europe were constructed with the Golden Mean in mind, and it was thought having these buildings in cities and working in them engendered a mood on the city and individual, a good mood.

A more contemporary attempt to bring nature's algorithms into life can be found in urban architecture and nature-inspired design, where you can find buildings modeled off the algorithms of nature: a branching canopy, a mushroom, the bark of a tree. Beijing's National Stadium comes to mind, aka the Bird's Nest. What differentiates these forms of algorithmic translation is that the algorithms in nature are to be applied in architecture, not the *modeling of data* in general.

The move to discern algorithms in nature so as then to implement back into nature or culture or the sciences should come as no surprise, as it is common in some circles to consider organisms themselves as "algorithmic," which is not too far off if one approaches the organism from a developmental perspective, as in yes, an organism is a biochemical algorithm. But is that all an organism is worth, or tells us about itself, or its final characteristic? Is that all that a dog is, a biological algorithm, which wags its tail at your very sight and gives you unconditional acceptance...it is just an algorithm? Aurelius spoke at length of the "well constituted" things of nature, noting their grace and attraction. Have we lost this quality of attunement to the organisms and mysteries of life in favor of their deep functionality? Arguably so, which is an indicator of how technology is not only informing what we know, but how we learn and what we find attractive to learn.

Lucy

Some animals can just keep growing, but it appears the power of intelligence is encountering a lot of growing pains. Not least of which in its desire to rule, it has had to wear a new mask. The new mask is a new self-definition, wherein intelligence becomes an operational procedure existing in nature, to the detrimental effect of reducing nature to a system of data exchange; in fact, the predominant method of prediction research is through the reduction of a cultural system or a brain to a series of functions. What does it say about a culture wherein they believe their intelligence is needed where it is not, such as in Smart Earth initiatives?

Lucy is the name anthropologists gave to one of our earliest ancestors—the movie *Lucy* (2014) strategically co-opted the name—and when you give a character name Lucy and specifically show the connection between the early human "prototype" and Johansson, also Lucy, the audience is spoon-fed a good couple of bowls of how techutopians views its historical contributions—as a fulfillment of history, as the ultimate path of humanity.

The film *Lucy* is not about the power of automation, but rather the power of drugs to open hitherto untapped regions of our brain and the same is true for *Limitless* (2011). However, the grand evolutionary narrative of *Lucy*—that it is humanity's destiny to activate more parts of our brain—is exactly the same narrative of techutopians, which unabashedly claim tech's "logical" goal is to integrate the sum total of all known data streams into a mind-software interface so we can be, like Lucy, fully immanent (present, equal, coterminous) to all that is, was and will be. Kurtzweil is explicit regarding the evolutionary link between biology and technology,

> Our ability to create models—virtual realities—in our brains, combined with our modest-looking thumbs, has been sufficient to usher in another form of evolution: technology. That development enabled the persistence of the accelerating pace that started with biological evolution. It may continue until the entire universe is at our fingertips.[8]

After successive openings of her brain to fuller and fuller usage, in the end Lucy becomes a supercomputer. If early proponents of automation thought it bold to claim the powers of technology would allow us to have more human joys, such as Wilde, it is mindful to note here the objectives have altered dramatically— the human joys are lost, or better, they are merged with a kind of informational bliss. In a nod to the inhuman future of interconnected, informational matrimony it needs to be stated that Lucy smiles progressively less and less throughout the movie the more she uses her brain.

The scientific method did change everything about the world when it solidified its observational and testing process in the sixteenth and seventeenth centuries, but the *extent* of influence does not equate to *quality* of influence—an abusive parent may be the biggest influence in a young child's life, but this in no way makes it the most beneficial influence. However, what one cannot argue is that our scientific, code-laden intelligence is the end product of evolution, *as if* there were a progressive development from burying our dead (which some scientists say is an initial sign of reflectivity), using tools and herding animals to corporate smart Earth initiatives. So, if there is no evolutionary story to be told about human intelligence—for example, more powerful yes, more effective no—then what is to be said about the "what is intelligence?" question?

What is important to the *what is intelligence?* question is not the answer, but what problem the solution serves. In our present case, the notion of intelligence as fulfilled in technology only serves the goals and ambitions of the latter.[9] Hence, indeterminate growth is just a quality of how we understand the trajectory of intelligence. It was always known intelligence had the ability for indeterminate growth, but now it has the environment, literally and figuratively.

Chapter 9

Terromatic Empathy

Man is but a reed, the most feeble
thing in nature, but he is a thinking reed.
Blaise Pascal

It is called *The Merge*. This is at least the most recent name for the fantasy. Michelangelo's *Sistine Chapel* and the languid finger of Adam reaching over to touch that of God is a version of the merge; so, too, is the universal archetype of the "green man," a personified nature-human being. Pan, the Greek half-man half-animal is a merge of sorts. The merge of our bodies or minds with something external to us is so universal in human history that it forms the core of every religion the Earth has witnessed — closeness to God, or Allah, or Brahman, or samsara, or truth, or nature. You get the point. It goes by other names, too — bliss, transcendence, ecstasy. It seems humankind has always fantasized about leaving our bodies, and technological societies, such as ours, are likewise seduced by the idea.

But, can we leave the Earth while still living on Earth? I do not mean, can we live in another place than Earth. That answer is obvious — of course we can. We probably will soon, at least with a few celestial colonies and such. I am asking — can we, today, exist in some ethereal computational sleep, a type of integrated virtual reality as envisioned by the tech community at large? We are talking about removing conscious life from our bodies, the twenty-first-century version of the merge. Elson Musk speaks for many when he writes: "The full-on-crazy version of the merge is we get our brains uploaded into the cloud. I'd love that. We need to level up humans, because our descendants will either conquer the galaxy or extinguish consciousness in the universe forever.

What a time to be alive!"[1]

Romantic Overlooks

So, if you find this *place* necessary for some reason, we must ask why? What is worth saving here? And why? This is not a question of environmentalism, but it is perhaps related. Rather, the question is about what purpose the Earth serves *for us* today since, though it might no longer be the case, we can all agree that for the entirety of human existence, with a small exception constituting the last 6000 years or so wherein large urban structures first appeared in ancient Mesopotamia, we have intimately evolved alongside its forests, villages, rivers and mountains. Given that Homo sapiens are 200,000 years in the making and complex urban cities are only 6000 years old, we have only lived apart from the rhythms of nature for 3 percent of the history of our existence as a species.

For 97 percent of human history we have evolved alongside natural habitats, and our bodies and minds have therefore been conditioned by our intimate existence with the Earth; yes, the last 6000 years may have altered how we form relationships and spend our time, but we have not grown another set of eyes or ears since Mesopotamia to better absorb the fast pace of civilization. In short, civilization may have applied a new color to the car, but our intimate life in natural landscapes created the car. It is in attending to the car we learn how to drive. While it is true gene editing and biotechnologies are soon going to be able to edit the car—making body parts, strong bodies, etc.—we will likely become that amateur mechanic who took apart their car in the garage only to discover, months later, they have no idea how the parts fit back together. If, or when, the car gets put back together, the drive will never be as smooth.

Over the long arc of history, one could reasonably make the argument that, assuming there was a time when people and the Earth lived in concert—mutual dependence and respect—

this relationship is on the rocks. It is no surprise megacities are altering our relationship with the natural world, a term used to describe the man-made art of civilization containing and entertaining people in excess of 10 million. In 1950 there were two megacities, in 1990 there were ten megacities and in 2014 there were twenty-eight. Tokyo takes the prize with 38 million inhabitants and the greater Tokyo metropolitan area measures about a third the size of the UK. This means the population of Tokyo has about as many inhabitants as the entire planet Earth did between the 2nd and 1st millennia BC. And, as of 2014, 54 percent of the world's population lives in urban areas—a number projected by the UN to be 66 percent by 2050. We need not lament these trends, either.

For millennia, human vitality depended on the Earth's vitality, or at least this was the *working assumption*. When a community performs a ritual in the hope of increasing the crop yield for the next year, they have placed their acts *between* two types of vitality—human and natural. For instance, the Native American rain dance was mean to bring rain and protect the harvest. Regardless of the ritual's efficacy, it was the working assumption the dance was necessary for the earth's health, which meant the people's health; likewise, more practical things such as nomadic movement or selective hunting ensured the human-nature relationship stayed strong, since it was mutually beneficial. And for a large proportion of early mythologies, there did not even exist the idea of a human apart from nature, but rather a shared vitality of which each part was simply a spoke in the cosmological wheel.

Since a shared vitality was necessary, empathy with the Earth was foundational and it was part of our evolution. It is still there, irrepressible, a voice we cannot snuff out as hard as we try. It is not like a tool in the box we are not using, but rather the body itself. Our orifices—ears, eyes, tongues, skin, and sensate fingertips—produce the emotions we have and are the product of thousands of years of having to listen, and adapt, to the Earth. Verily, though we work on it, the work is largely being done *to us*. We are still *its* actors, still working for it, on it and from it. It will be this way until generation after generation breed in celestial colonies on lunar landscapes, when those minds are half human, half machine and are reassembled so many times the faint trace of existing on planet Earth is gone. However, one imagines even then, even there, instinct will still attach itself to time and space, and the shapes of the new home will eventually alter their psyches as much as Earth did ours. Their minds will become lunar and the landscape will write on their bodies as the Earth has ours. To be a body might just mean, by default, we are bound to it, our imagination structured by it, our religions and philosophies born from it. What would it even mean to *imagine* when we are digital minds on a cloud without reference to material things?

Terromatic Empathy

Famed biologist and thinker Edward Wilson proposed a controversial biological "law" of sorts and it is as simple as it is hard to prove. Wilson called it *biophilia* and coined it to describe "the connections human beings subconsciously seek with the rest of life"; Wilson also described it as our "innate tendency to focus on life and lifelike processes."[2] Technically, the term means the love of life, but its value is not in its definition but rather as a description of fundamental human behavior.[3] Estrangement from the rhythms of the natural are a real thing, but flowing underneath estrangement is a *terromatic empathy* that can only be broken if we leave our bodies altogether, which seems more science fiction than science. Terromatic empathy is a co-feeling, a care for natural vitality and a source of stress when that vitality is disrupted. If biophilia is the phenomenon of a human/ nature kinship, terromatic empathy is the expression *in us* of the phenomenon.

Where is *terromatic empathy* expressed? In flowerpots on your mother's front porch, Central Park, Yosemite, the plant on your office desk, that feeling when we see a river choking in plastic bags, a poor child playing in garbage, your walk during the lunch hour, the fact we enjoy grass or trees in our yards, working professionals who come home to the favorite part of their day — seeing their dog or cat. There are biophilic trends in office decor, feng shui and architecture whose modeling systems use patterns (algorithms) found in nature, such as seashell spirals or other esoteric patterns. Biophilia even shows up in technology. Sue Thomas acutely asks, "Why are there so many nature metaphors — clouds, rivers, streams, viruses, and bugs – in the language of the internet? Why do we adorn our screens with exotic images of forests, waterfalls, animals and beaches?"[4]

One of the great cruxes of biophilic theory has been to "prove" how the world around us has played a part in our psychological development. The field of research attempting to exhume the way the natural world has influenced the how and what of our physiological constitution can be found in evolutionary psychology, ethics, environmental psychology, neurology and anthropology, among others. It is a growing field of research for those interested and should dovetail nicely with the question of "why" to the other fields of experimentations trying to understand the "how" of our bodies and minds.

Still, despite the common sense claim that the environment in which humans were cultivated over hundreds of thousands of years would have an effect on their brains, finding exact correlations, as in cause and effect, proves elusive. Work done in the realm of phobias gets very close to pinpointing the effect of developing alongside nature, in this instance, snakes. For example, biological researcher Kenneth Kendler wonders why most people have extremely overexaggerated responses to snakes when in reality snakes are not a threat to us.[5] Oddly, we can hold a gun just fine, but holding even a harmless snake activates something primeval. For Kendler, our reactions have been conditioned by thousands of years of living outside with snakes, in which snakes could be dangerous. Biophilia is expressed, in this example, in phobias now hardwired into our genes.[6] But that is just snakes. How to talk about the influence of a river on a society living on its banks for thousands of years? Remarkably, some have attempted, concluding that peoples who have lived by calm, large and predictable rivers tend to characterize their gods as well behaved, patient and understanding, as opposed to peoples living near rivers that flood and devastate, wherein the latter's gods are characterized as irrational, prone to violence and uncompromising. What of the shape, speed or character of water in the psyches of coastal peoples?

The most watertight argument, however, to prove the Earth has been essential to our development is that we are an organism and organisms are of the Earth. When French philosopher Blaise Pascal wrote, "Man is but a reed, the most feeble thing in nature, but he is a thinking reed," he hit on a simple truth. While our existence in the flesh does not prove we are *consciously* attracted to life and lifelike processes, it proves we have evolved from the rhythms of life much as a canyon has evolved with the rain—it proves our bodies are attracted to the Earth insofar as all of its joys, to date, are generated from it. And mankind seeks joy. Yes, a daughter or son can be born and later become so fully estranged from their parents they care for them not one iota, and yet, the metaphor is only fleeting and superficial, as the structure of our instinct and body remains wedded to our primitive development. Insofar as we are, en masse, awake during the day and asleep at night, fear the darkness, worry about shelter or hunger, believe groups are stronger than individuals, court mates—insofar as all these things ring true, we are products of the Earth and as products of the Earth, our relation to it will forever elide the narrowness of conscious thought for the simple reason that conscious thought is itself a product of a human/nature interface.

But what does any of this have to do with algorithms?

Blood Expansion

Nature loves to hide.
Heraclitus

Attempting to describe "the connections human beings subconsciously seek with the rest of life," what should strike one as odd about biophilia's definition is *connection*. Why is that odd? Well, it just so happens that "connectivity" is arguably the *raison d'être* for the proliferation of algorithm-fueled automation. Not just connectivity to the Internet, but IOT, ubiquitous computing, smart cities, smart earth, etc., *all but variations of connectivity*: devices to devices, people to devices, devices to nodes, homes to nodes to devices and, you guessed it, algorithms are further refining how the connections become more meaningful, targeted and intimate. The Internet of things *connects* us to our environment; dating apps *connect* us more efficiently and ideally more precisely; we get *connected* to music we love; our friends *connect* with us, with what we like and share and promote. On the other side of things, our homes *connect* with us and our devices literally *connect* with us so we can, in turn, be more *connected* to our bodies, or our children and communications. Augmented reality *connects* us with virtual worlds, our fantasies and our fears; virtual reality *connects* us with entirely new worlds. Smart-earth campaigns *connect* the Earth to itself and to us, and smart cities *connect* infrastructures to each other and to us.

While it appears connectivity is not the goal, but the enabler, this is not the case. Just as the automation ethos is an unquestioned value, so, too, is connectivity. Connectivity is the goal and ancillary benefits are just that. To be more connected is simply a greater form of being. A personalized world via connectivity is the largest, most expansive trend in technology, and this is likely to be the case for the foreseeable future. Some thinkers, notably Harari, cite *dataism* as the predominant religion of the future and the current ideology of tech centers around the world.[7] That much is undeniable for a few, but for the rest of us data is not important. Data is technical and very few of us go about our day thinking about data. We think about being connected. Data is simply what makes connectivity possible and connectivity is a form of life, albeit one existing alongside, or, within —*what preposition is it?* — traditional life, pre-digital life.

As a result of connectivity, our world is becoming increasingly animistic. As Betti Marenko and others have observed, animism defined as the "idea that objects and other nonhuman entities possess a soul, life force, and qualities of personhood."[8] More precisely, our world is techo-animistic.[9] Thanks to IOT, everything is now alive or soon will be: moving thanks to its own technological agency, adapting, feeding, morphing, routing, catering, adjusting, responding. Google's self-driving car, which is really a computer, can now be considered by the US National Highway Traffic Safety Administration to be a "driver." Likewise, will systems in the future be given individual rights, as we are? Obviously, we are near the question of if androids could be considered persons.

The specter of a vitalistic, techno-animistic world is upon us, promising to be even more alive than life without it. On the one hand there is connectivity, arguably the newest motivator in human life, and on the other hand is the oldest, biophilia. When put together, a shocking portrait emerges of human energies working in remarkable *unison*, not antagonism. As it turns out, technological connectivity and natural biophilia are the same force, the result of the same impulse. To put it into a question— what instinct, reflex, love, drive or libido package is motivating us to want to create an alternate, virtual form of "life" within *and* on the already existing textures of life? The force behind connectivity and automation, it turns out, is biophilia. We are witnessing a *vitality compensation*.

Tech visionaries are some of the biggest dreamers on the world stage and they need to be commended as such. It should come as no surprise, therefore, that we are using technology to make the world *more alive*—giving life—in an era where civilization has dislocated us from natural vitality. Whether we know it or not, we are putting one vitality in place of a lost one. It was the Industrial Revolution and the industries of extraction first exacerbating the decline of the Earth's health, and considering technology could have gone in a lot of directions, the fact it is directing its tools to more life via interconnectivity is not by chance. In other words, we are infusing our environment with a techno-animistic intelligence at the exact same time in history that our culture, physiology and environment is finding it harder and harder to be intelligent by itself. What Bart Simpson said of alcohol applies to connectivity: "it is the problem and solution at once."

The Trojan App

But here is the question—if biophilia is an innate physiological motivator that is *outward* facing, drawing us to forms of life as they exist outside our bodies, then is the push for more connectivity with the outside world, and the giving of objects a quasi-life, just an updated expression of biophilia? Is the most powerful force within the most powerful industry simply a repurposed ancient drive? That underneath complexity and advanced civilization lies basic human drives, such as primitive fears and motivations, should come as no surprise to any student of humanity, politics or religion.

On this accounting, however, an Earth-inspired drive guiding technologies *ought* to be a therapeutic force, a corrective, a welcome addition to populations living largely in urban, built-up environments, properly removed from nature's cycles. We should welcome techno-animism as a pseudo-replication of a type of environment we developed alongside for tens of thousands of years. Yet, there is a problem in the thesis—as biophilia developed and expressed itself in us as terromatic empathy, the Earth *was* a vibrant, functioning holistic organism, which it has been for the majority of human development. The problem is—it is not. We have two varieties of connectivity, but they are oppositional in nature. We are trying to connect with an Earth whose living systems are in disrepair and so, as a result, we are sowing the seeds for our own disrepair.

Contemporary biophilia (techno-animism) is maladaptive. Why maladaptive? Terromatic empathy, as the golden thread connecting us to the Earth, is guiding the process of biophilia. Whereas we *were* working in unison with a vibrant Earth and became more vibrant as a result of our labor, we are still acting out of the connectivity impulse, but with a significant difference—the drive is corrupted since the Earth's self-moving "intelligence" is fragmented. In short, we are cultivating our own degenerative nonvitality—a fancy phrase for uselessness, disengagement, etc.—under the guise of a supposed generative vitality (techno-animism).

While adaptation is an essential act of our physiology, terromatic empathy is forcing us into a non-caloric, neo-hibernation of sorts, a shutting down of our bodies and minds that, ironically, mimics the Earth's trend towards non-vitality. We can concretely envision the future and it is advertised as such by tech companies as a body lacking motion, with its interior world being the main source of life. It started with digital products, then moved to the big screen, then our living rooms, then our work, then in our pockets, then onto our faces as virtual reality, then maybe a type of integrated reality, then, if the story proves trustworthy, we will move entirely into the digital world, consciousness and all. It is all a process of going inward. We are slowing down all parts of our physiological systems just as, no coincidence here, the Earth is slowing down. The phenomenon of algorithms fueling automation and the subsequent removal of ourselves in the process of decision-making, cultural participation and stewardship of the Earth is an expression of a going inward, not a going outward (classic biophilia). To this end, techno-animism is a Trojan horse, a modern-day rain dance, but with little chance of success.

Chapter 10

Restore Point

Everyone gets old and most people lose the ability to be rational beings as the end of life nears. Age happens. It is often ugly, too. If you have ever had to deal with aging parents or grandparents, strange contingencies need to be put in place and we all have to prepare for a strange thing—who will make decisions for them when they are no longer able to? The truth of the matter is that most, with few exceptions, are able to make decisions up until the day they die. Making a decision is not the problem. The problem is—they make bad decisions, such as refusing to take a tiny innocuous pill that will save their life. The whole process of preparing for end-of-life care is at once logistically sobering with all the involved legality and emotionally it is a wildfire, as someone is effectively acknowledging the time when they will be, but will not quite be.

Computer engineers acknowledge a similar phenomenon with "restore points" on your computer, which exist should your computer become infected and is thereby unable, like an aging relative, to make a good decision. A restore point is a place where your computer can go in the *future* should something happen *today*—like a point in the past when the computer was happy and functioning normally—but it will not be able to do it itself, since at that point it cannot make a good decision. The user has to implement the restore point. Sons and daughters are like rational restore points for aging parents, ensuring they approach Elysium with some dignity, since that is really what it is about, dignity. The thoughts presented here are restore points for a humanity likely, but not definitively, in the process of losing the rationality it took so long to create.

∞

This book was not really about algorithms, AI, integrated reality or the internet of things. This book was about you and I, children, creativity, instincts, the future, romance, great sex, fantasy, boredom, depression, our souls.

First, we have humanity—in the mist of finding its labor and intellectual powers outsourced, rendering our decision-making skills useless. Algorithms are intervening in all places in our lives, between us and others, us and our minds, us and our bodies, us and the Earth, us and the things we rely on to tell us about ourselves. Not a conspiracy, not a boardroom secret, but simply teams of software engineers trying to make their algorithms as efficient as possible. Studying how outsourcing jobs to India affects US workers is easy. Studying how outsourcing all of these emotional or intelligence tasks is not work for the scientist, but for the philosopher. This question cannot traffic in data and studies and results, but must confront, head on, questions going to the core of human life, vitality and value. Even if the jobless future does not come to pass, the nature of the work will be changed forever. With only a select few positions to remain unharmed, a large percentage of future work will be systems management, monitor-ish and techno-bureaucratic. While this should not be lamented *per se*, the fact that humans become increasingly disengaged and lose awareness with their surroundings in automated environments should be an early warning sign to the volume of joy the future will produce.

But the future plows inescapably forward and the automation ethos, already well entrenched in our psyches, will fuel our desire to free up time in our frenetic lives so we can...write poetry?... lollygag under a dogwood?...spend quality time with friends?... well, we do not quite know. The historical premise, and promise, of automation was its liberatory potential, and its nectar was advertised well before our era. When our work is done for us, we can dream more, have more leisure, write, create. But reality is a cold bath and we are not such idealistic creatures who default to cultivating bliss when the yoke of work is unharnessed.

As societies are becoming more automated, we are working more hours and when we should be the ones creating new objects to move culture forward, we are outsourcing that too, such that if, or when, the time comes when we actually do have free time—some say it and the basic income are coming—we will not even value the creative process. The governments of the future, therefore, will become more adept at managing the coming health crisis of boredom and it will struggle to manage the cache of energies release by boredom—disengagement, cruelty, irrationality, feelings of uselessness. The most difficult hurdle in technology is not technological, but in figuring out how to shoehorn its creators, us, into *its* workflow to keep it functioning. Unfortunately, nearly all scenarios about the future, even conservative ones, must contend with wide-scale human uselessness.

Next, we have culture, which is beginning to be populated with objects of whose origin we are unsure. A pool of circulated objects and events increasingly governed by technological agency. Is that photograph I like computer generated? What about that song? Or novel? Music, film scripts, aside from currently being "vetted" by software, will soon be created by software and they will be seamlessly integrated into our cultural landscape. Everywhere, culture will be catered to us, personalized at the same time as it is anonymously outsourced. The ability to discern, nay, even the practicality of developing a sense of taste, shall go the way of auricular muscles, which allow animals to swivel their ears to hear for predators. We have them too, but they serve no purpose; they are a relic of a bygone era.

If culture is a collectively woven fabric, which should work *for us*, we should caution that its purpose will no longer be for us if on this path, but will create the conditions for the thing generating it. This is a natural consequence, as much as the beaver makes the pond for him. Culture will more closely resemble a controlled environment and as a result, cultural options will decline (the "you loop" is but an example), and with that our fields of action; when this occurs, the range of emotions offered to us will shrink and we will experience less freedom as a result.

Ironically, the introduction of intelligent machines might increase a slice of our physiological behavior—for example, human intelligence—but in large measure it will make us less free. Human thought and consciousness are much more than our intelligence. Automation is the first of its kind in a new type of creative object—it is the subtraction of experience, not an intensification of experience. If culture is the product of interactions whereby we embed shards of humanity into those products, such as the mood of a literary character or the blues riff stopping you in your tracks, then the missed opportunity and ensuing degradation is far from theoretical but wholly imminent.

Alongside automation is predictability. A growing body of research is being directed at predicting human thought, behavior and movement. Soon, broader theories of humanity will merge with complex software able to work in mysterious ways, thereby producing a level of complexity we know not what. Our era has been called the Age of Entanglement, because no one understands what is really happening with the majority of things we use, and yet complexity is only going to increase. Typical responses to complex environments include stalling, frustration, despondency and helplessness. Ultimately, disengagement arrives and then boredom.

Finally, we have nature—increasingly under the yoke of data. Smart Earth campaigns are building a layer of intelligence on and inside the innermost processes of nature, more often than not to save it. Trees with Twitter accounts, thousands of sensors everywhere. We are only, realistically, in the data-capture phase of natural data but, just as we were in the early 2000s with the gathering of consumer information via big data tracking, it will not take long until the intervening phase begins. The future will bring designer ecosystems, the modification and reintroduction of species, and the saturation of ecology with monitorization. The latter is undoubtedly an old trend, begun when humans first started tracking animals, then tracking became husbandry then breeding. We are building a layer of intelligence on top of natural systems, effectively making it impossible for nature to exercise its own (natural) intelligence. The loss of a functioning natural intelligence will equally affect us, as co-inhabitants of our Earth.

Biophilia, as the instinctual desire to connect with life, is remarkably fueling the technological products and trends of our world. We want to be connected. The more connected the better. *Being connected* is a state of being, just as being moral is a heightened state of being for some Christians, or being mindful is a preferable state of being for some Buddhists. However, in the push to surround ourselves with living objects (techno-animism), we are merely duplicating a lifeless Earth thanks to terromatic empathy, so the ultimate dream of tech enthusiasts of a joy-filled future of human-computer interfaces—aka the merge—necessarily fails. It is not a joyless future *per se*, unless you think a computer can feel joy. Since we are replicating a dying Earth, we are also cultivating a sense of "death" in our lives. Uselessness, and all it brings with it, is the clearest expression of this death.

The Thread

What is the thread between nature, culture and humanity? Increasing uselessness. To be "of use" means to have utility, play a role and, in said role, construct just about everything we take for granted—political theories, art objects, memes, great food or beautiful tree-lined streets. To be "of use" is not about feeling unique or special. While certain sectors of small populations in history have had to question their utility, the scale stops the analogy in its tracks. Uselessness in the sense not being necessary for basic and complex needs—at the personal and cultural level—is, while in no sense foregone, more likely than not.

The effects of uselessness regarding culture are pointed— *culture need no longer serve us*. Once this becomes apparent, culture becomes useless as a tool to do its traditional duty, since, quite literally, it is no longer *for us*. Humanity and culture are therefore triangulated, entangled, in the cycle of uselessness: the more estranged we become from increasingly interacting with inhuman cultural objects, the more useless we become to culture, and to ourselves, and the more we lose our soul. The more we allow technologies to reduce our decision-making and consciousness abilities, the more disengaged we become, which renders our desire to contribute to culture less.

Fracking is a well-known term in the parlance of geopolitics and it is infamous for its externalizations (creating hidden costs elsewhere), such as polluting drinking water via underground aquifers traveling deep beneath the earth's surface in spite of claims of its airtight procedures. Analogously, we have a case of *biofracking,* wherein in the laying of "pipe" into our innermost being we activate the *black swan*—the thing that cannot be predicted, but which will have the most consequence. In the case of natural gas extraction from shale, it is not hard to predict some gas is going to pollute drinking water but, in the case of biofracking, the stakes are much higher and the underground aquifers essentially unknowable. Every generation frets about the future, as the future is basically a mystery, but any good coach or parent knows that to stem some behaviors in the future, and to avoid a total disaster, the work begins today.

Robotic engineers and the surrounding literature debate at length the uncanny valley hypothesis, which states robots that look "too human" are off-putting to us, even repulsive.[1] What about the uncanny valley effect when we behold or participate in things like the natural world or the stuff of culture, the latter two becoming increasingly programmatic? There is no research into how the uncanny valley effect will operate when it is not just inhuman robots we are encountering. Typically, stopgaps are implemented in systems as negative feedback mechanisms to prevent intensification—such as a thermostat—but as of yet we do not have any. The effects of uselessness on humanity are pointed—disengagement, cynicism, despondency and boredom. We will see these energies deployed in our political spheres first, as "the vote" is a primary means of venting frustration. Alongside this will come antipathy, resistance, irrational violence and cruelty; not civil or global war, but circulating negative energies in the most "peaceful" times in history.

The more our bodies are mined for its biological data and the less we rely on our internal biological feedback systems, the less we feel the need to have bodies, thereby accelerating the divorce of consciousness from intelligence. Biophilia and its accompanying feature, terromatic empathy, originally put our physiology in a vital and undeniable relation with the Earth, regardless of how it was managed. However, the drive once engendering vitality is now wedded to a non-vitalistic Earth and the once-adaptive drive can now be considered mal-adaptive. In many ways, we need to rewire our instinct, if this is possible..

The more culture stops working for us, stops feeding us and teaching us the rules of human engagement, the more likely we are to destroy it—it becomes useless. We can all understand what it means not to have cultural progress from an ethical or political angle—WWII comes to mind—but still, the metrics are nebulous as to how to judge the vitality of a creative culture; the ideology of cultural relativity prevents rigorous thought in this manner. The question of *who we want to be* needs to be asked in a more pressing manner and decisions need to be made, otherwise the decision will be made for us. A restore point is needed.

Even if we cannot answer the question of what we want to be, which we can't, we can agree on our value as participating members within ourselves (mastering our bodies), and as valuable in the economic, emotional and creative economies surrounding us. A sure point of criticism is that this perspective—the one of human value—is still a humanism and that in 50 years the whole humanist project will be irrelevant, since humans will no longer be the "center" of our world. This point could be potentially true from an ideological standpoint, but not an operational one. Unless the AI apocalypse does come true and we are taken over by robots that think humans are a nuisance, which even the most hardened tech enthusiasts deem extremely unlikely, then our power remains sovereign. [2] Just as every dollar spent was a vote in a paper economy, every act, every move and every click is a vote in the era of big data.

The most sinister of cultural algorithmic creators can be dismantled in a manner of weeks if we simply stop using them, yet, the use of big data in the realms of finance or creativity will likely go unchecked; the latter will not be noticed, because we are never told when we are consuming an automated photograph or story. As Angela Merkel demanded that social media algorithms be made public, on account of being politically dangerous, likewise we need to develop a truth-in-algorithm tag for the digital era, something to communicate to us when and how a thing was created. If novels, paintings and large sectors of the creative industries will be awash in algorithmic objects and our innermost citadel becomes atrophied by the myopia of algorithmic options, the likely result will be a net-depreciation in the value and quantity of self-reflection, utility, freedom, intensity of experience and joy, since software can experience none of these things.

References

Introduction

1) Agerholm, H. Angela Merkel says Internet search engines are "distorting perception" and algorithms should be revealed. BBC News, October 2016. One of the problems in writing about the costs of automation—in other words, algorithms—is studies are hard to come by. This point cannot be underscored enough. Actually, the research is virtually nonexistent. It is one thing to track the number of laid off workers in Chinese manufacturing who lost their jobs to robots using sophisticated algorithms to automate processes, but it is another to track how Pandora, fitness apps, automated creativity and Waze—to name the big ones—are undermining our ability to be willfully guided, vital human beings, to be discerning bodies of taste and producers of culture. The former gives itself to statistics, research models, facts even, while the latter must be addressed in a different manner, theoretically and philosophically; however, this shall not deter us, since there are no scientific facts to support the claim that art is definitely a good thing for culture, and yet we do not need proof to believe it, to know it. The greatest threat to human civilization in the coming years, therefore, will not come from algorithms per se, but in how they will render human consciousness and our cultures useless.

2) Guynn, J. (March 2017). 'Facebook begins flagging "disputed" (fake) news'. *USA Today*. Available from: <http://www.usatoday.com/story/tech/news/2017/03/06/facebook-begins-flagging-disputed-fake-news/98804948/> [accessed 19.04.17]

3) Harari, Y. (Interview recorded March 2015). 'Death is Optional'. Radio interview with Daniel Kahneman for the

Edge. Available from: <https://www.edge.org/conversation/ yuval_noah_harari-daniel_kahneman-death-is-optional> [accessed 19.04.17]

4) Brynjolfsson, Erik, and McAfee, A. (2014). *The Second Machine Age: Work, Progress, and Prosperity in a Time of Brilliant Technologies*. New York: W. W. Norton & Company.

5) On another front, modern telecommunications are just too good and global geo-politics too fine-tuned to let war and regional disputes spiral into global chaos. The big global economies are too interdependent for a mutual self-destruction, since the physical effects of conflict are often shorter-lived than damaged economies, castrated development and dearth of investment. This is, in essence, what Thomas Friedman means by his "Big Mac" theory of war: "No two countries that both have a McDonald's have ever fought a war against each other (Friedman, Thomas L. (December 1996). 'Foreign Affairs Big Mac I'. *The New York Times*). Economies that are fully industrialized do not go to war because their economies, the real lifeblood of a country, are too intertwined. Moreover, in a sentiment seemingly repellant to common sense, according to Pulitzer prize-winning author Steven Pinker we live in the most peaceful time in our species' history. That is likely to continue (Woolf, C. (September 2014). 'The world is actually becoming more peaceful — believe it or not', according to Pinker in a 7/16/2016 NPR.org interview). Our biggest threats are not hunger, war or climate change.

6) Osbourne, Michael A., and Frey, C. (September 2013). 'The Future of Employment: How Susceptible are Jobs to Computerisation'. Oxford University. Available from: <http://www.oxfordmartin.ox.ac.uk/downloads/academic/ The_Future_of_Employment.pdf> [accessed 19.04.17]

7) Economist.com. (December 2014) 'In search of lost time: Why is everyone so busy?' *The Economist* [online].

Available from: <http://www.economist.com/news/chris tmas-specials/21636612-time-poverty-problem-partly-perception-and-partly-distribution-why> [accessed 19.04.1 7]

8) Graeber, D. (August 2013). 'On the Phenomenon of Bullshit Jobs'. *Strike Magazine* [online]. Available from: <http:// strikemag.org/bullshit-jobs/> [accessed 19.04.17]

Chapter 1

1) Fuller, Matthew, ed. (2008). *Software Studies: A Lexicon.* Cambridge: The MIT Press.
2) Ibid., p. 16.
3) Ibid., p. 15.
4) Cormen, T. (2013). *Algorithms Unlocked.* Cambridge: The MIT Press.
5) Ibid., p. 1.
6) Executive Office of the President, President's Council of Advisors on Science and Technology. (December 2010). *Report to the President and Congress—Designing a Digital Future: Federally Funded Research and Development in Networking and Information Technology.* [pdf] Available at https://www.cis.upenn.edu/~mkearns/papers/nitrd.pdf [accessed 19.04.17]
7) Nickerson, D., & Rogers, T. (2014). 'Political Campaigns and Big Data'. *Journal of Economic Perspectives*, 28 (2), 51-74.
8) MIT Technology Review. (January 2016). 'How an AI Algorithm Learned to Write Political Speeches'. Available from: <https://www.technologyreview.com/s/545606/how -an-ai-algorithm-learned-to-write-political-speeches/> [accessed 19.04.17]
9) Miklos, B. (November 2015). *Computer, Respond to this email: Introducing Smart Reply in Inbox by Gmail* [Blog] Google. Available from: <https://blog.google/products/gmail/ computer-respond-to-this-email/> [accessed 19.04.17]

10) Epagogix.com. *Helping Business Leaders make Big Decisions.* [online] Available from: <http://www.epagogix.com/> [accessed 19.04.17]

11) MIT Technology Review. (November 2014). 'Machine-Learning Algorithm Ranks the World's Most Notable Authors'. [online] Available from: <https://www. technologyreview.com/s/532591/machine-learning-algorithm-ranks-the-worlds-most-notable-authors/> [accessed 19.04.17]

12) Hu, E. (December 2013). Microsoft Not Developing A Bra To Stop Overeating, After All. [online] Npr. org. Available from: <http://www.npr.org/sections/ alltechconsidered/2013/12/10/249963461/microsoft-not-developing-a-bra-to-stop-overeating-after-all> [accessed 19.04.17]

13) Studioroosegaarde.net. See Studio Roosegaarde's Intimacy 2.0 advertising on their website. Available from: <https:// studioroosegaarde.net/project/intimacy-2-0/> [accessed 19. 04.17]

14) Lundin, E. (November 2016). Could an algorithm replace the pill? *The Guardian.* Available from: <https://www. theguardian.com/lifeandstyle/2016/nov/07/natural-cycles-fertility-app-algorithm-replace-pill-contraception> [accessed 19.04.17]

15) Borreli, L. (March 2014). 'E-Spot' Implant Helps Women Climax At The Push Of A Button: Could These Implantable Devices Treat Orgasmic Dysfunction? [online] MedicalDaily. com. Available from: <http://www.medicaldaily.com/e-spot-implant-helps-women-climax-push-button-could-these-implantable-devices-treat-orgasmic-270697> [accessed 19.04.17]

16) Weissmann, J. (June 2012). iLawyer: What Happens When Computers Replace Attorneys? *The Atlantic.* Available from: <https://www.theatlantic.com/business/archive/2012/06/

ilawyer-what-happens-when-computers-replace-attorneys/258688/> [accessed 19.04.17]

17) Xiao, B., Imel, Z. E., Georgiou, P. G., Atkins, D. C., and Narayanan, S. S. (2015). Rate My Therapist: Automated Detection of Empathy in Drug and Alcohol Counseling via Speech and Language Processing. *PLoS ONE* 10(12): e0143055. doi:10.1371/journal.pone.0143055 [accessed 19. 04.17]

18) Shepardson, D., Lienert, P. (February 2016). Exclusive: In boost to self-driving cars, U.S. tells Google computers can qualify as drivers. [online] Reuters. Available from: <http://www.reuters.com/article/us-alphabet-autos-selfdriving-exclusive-idUSKCN0VJ00H> [accessed 19.04.17]

19) Berk, R. A., Sorenson, S. B., and Barnes, G. (2016). Forecasting Domestic Violence: A Machine Learning Approach to Help Inform Arraignment Decisions. *Journal of Empirical Legal Studies*, 13: 94–115. doi:10.1111/jels.12098 [accessed 19.04.17]

20) Ibid.

21) Schmidt, E. (January 2015). As cited in Google Chairman: 'The Internet Will Disappear,' by Dave Smith. [online] Business Insider. Available from: <http://www.businessinsider.com/google-chief-eric-schmidt-the-internet-will-disappear-2015-1> [accessed 19.04.17]

Chapter 2

1) Harder, D. S. (October 1954). Automation. *Michigan Technic*.

2) Wilde, O. (1900). *Miscellaneous Aphorisms: The Soul of Man*. New York: Dossier Press.

3) Intel 3Q results dip but beat Street forecasts. Associated Press. October 2015. Available from: <http://www.huffingtonpost.com/huff-wires/20151013/us-earns-intel> [accessed 19.04.17]

4) Tesla, N. (February 1937). Master of Lightning: A Machine to End War. *Liberty Magazine*.

5) Keynes, J. M. (1963). *Essays in Persuasion*. In: Economic Possibilities for our Grandchildren. New York: W. W. Norton & Co.

6) Kurzweil, R. (2005). *The Singularity is Near: When Humans Transcend Biology*. New York: Penguin Books.

7) Marx, K., ed. by John Elster, (1986). *Karl Marx: A Reader*. Cambridge, UK: Cambridge University Press. p. 160.

8) Marx, Karl, ed. by Robert Tucker (1978). Marx-Engels Reader, 2nd edition. New York: W. W. Norton and Company. p. 286.

9) Ibid., 441.

10) Gates, Bill. Gates participated in a Reddit Ask Me Anything. January 2015. Thread can be found at: https://www.reddit.com/r/IAmA/comments/2tzjp7/hi_reddit_im_bill_gates_and_im_back_for_my_third/ [accessed 19.04.17]

11) Keynes, John Maynard (1963). Economic Possibilities for our Grandchildren. In: *Essays in Persuasion*. New York: W. W. Norton & Co.

12) Russell, Bertrand. *In Praise of Idleness and Other Essays*. New York: Routledge; 2006.

13) Ibid.

14) Ibid.

15) Carr, Nicholas (2014). *The Glass Cage: How Our Computers Are Changing Us*. New York: W.W. Norton and Company.

Chapter 3

1) Galilei, Galileo. Il Saggiatore. Rome (1623). See also The Assayer in *Discoveries and Opinions of Galileo*. New York: Doubleday and Company; 1957. Pgs 237-38.

2) Pythagoras (2009). See M. Brake's *Revolution in Science: How Galileo and Darwin Changed Our World*. New York: Palgrave Macmillan.

3) Nietzsche, Friedrich (1974). *The Gay Science*. New York: Vintage Books.

4) Russell, Bertrand (2006). *In Praise of Idleness and Other*

Essays. New York: Routledge.

5) Simon, Herbert (1971). Quote from a lecture. Designing Organizations for an Information-Rich World. In: *Computers, communications, and the public interest*. Baltimore, MD: The Johns Hopkins Press.

6) Rosenblat, Alex and Kneese, Tamara and Boyd, Danah. Understanding Intelligent Systems. October 8, 2014. *Open Society Foundations' Future of Work Commissioned Research Papers 2014: Data-Centric Technological Development in the Workforce*. Available at SSRN: https://ssrn.com/abstract=2537535 or http://dx.doi.org/10.2139/ssrn.2537535 [accessed 19.04.17]

7) Gillespie, Tarleton (2014). The Relevance of Algorithms. In: *Media Technologies*, ed. Tarleton Gillespie, Pablo Boczkowski, and Kirsten Foot. Cambridge, MA: MIT Press.

8) Stark, Miriam, Brenda Bowser and Lee Horne, eds (2008). *Cultural Transmission and Material Culture: Breaking Down Boundaries*. Tucson, Arizona: The University of Arizona Press.

9) Angela Merkel wants Facebook and Google's secrets revealed. *BBC News*. October 2016. Available from: http://www.bbc.com/news/technology-37798762 [accessed 19.04.17]

10) Madrigal, Alexis. Take the Data Out of Dating. *The Atlantic*. December 2010. Available from: https://www.theatlantic.com/magazine/archive/2010/12/take-the-data-out-of-dating/308299/ [accessed 19.04.17]

11) Pariser, Eli (2011). *The Filter Bubble: What the internet is hiding from you*. New York: The Penguin Press.

12) De Certeau, Michel (2011). *The Practice of Everyday Life*. Los Angeles: University of California Press.

Chapter 4

1) Gillespie, Tarleton (2014). The Relevance of Algorithms. In:

Media Technologies, ed. Tarleton Gillespie, Pablo Boczkowski, and Kirsten Foot. Cambridge, MA: MIT Press.

2) Striphas, Ted. Interview titled. Algorithmic culture. Culture now has two audiences: people and machines. *Medium. com*. April 2013. Available from: https://medium.com/futurists-views/algorithmic-culture-culture-now-has-two-audiences-people-and-machines-2bdaa404f643 [accessed 19.04.17]

3) Ibid.

4) Samothrakis S, Fasli M (2015) Emotional Sentence Annotation Helps Predict Fiction Genre. PLoS ONE 10(11): e0141922. doi:10.1371/journal.pone.0141922 [accessed 19. 04.17]

5) Eisner, Elliot (2000). Ten Lessons the Arts Teach. In: Learning and the Arts: Crossing Boundaries. Proceedings from an invitational meeting for education, arts and youth funders. January 2000, Los Angeles. Available from: https://www.giarts.org/sites/default/files/learning-and-the-arts-crossing-boundaries.pdf [accessed 19.04.17]

6) Gillespie, 2014. Also see Langdon Winner's "Modern Technology: Problem or Opportunity?," in *Daedalus*, Vol. 109, No. 1, (Winter, 1980), pp. 121-136.

7) Gillespie, Tarleton (2014).

II. Culture

Chapter 5

1) Saint-Exupery, Antoine de (1992). *Wind, Sand, and Stars*. New York: Harcourt Brace Modern Classics.

2) Campbell, Tom. The End of the Creative Class. March 2013. The original post was taken down, but his quote can be found here: https://www.dezeen.com/2013/03/11/machines-will-make-workers-in-creative-industries-redundant-tom-campbell/ [accessed 19.04.17]

3) Warhol, Andy (2004). What is pop art? Answers from 8 painters, part 1. In: *I'll Be Your Mirror: The Selected Andy Warhol Interviews*, ed. Kenneth Goldsmith. New York: Carroll and Graf Publishers.

4) Steiner, Christopher. Innovation is a social issue. *The European*. July 2014. Available from: http://www. theeuropean-magazine.com/christoper-steiner/7226-algorithms-and-the-future-of-work [accessed 19.04.17]

5) Sherwin, Adam. The perfect pop song? It's the algorithm, not the rhythm, that counts. Independent. July 2012. Available from: http://www.independent.co.uk/news/ science/the-perfect-pop-song-its-the-algorithm-not-the-rhythm-that-counts-7962590.html [accessed 19.04.17]

6) Lebrecht, Sophie. As cited by Jonathan Shieber in an interview. Quote was posted by Techcrunch.com on January 2015. Available at: https://techcrunch.com/gallery/here-are-the-best-6-images-from-ces-according-to-neon-labs/ [accessed 19.04.17]

7) PhiMatrix. See their website at Phimatrix.com.

8) Hotz, Robert. What's Hot in the Art World? Algorithms. *The Wall Street Journal*. May 2015. Available from: https:// www.wsj.com/articles/whats-hot-in-the-art-world-algorithms-1432687554 [accessed 19.04.17]

9) MacCormick, John (2013). *Nine Algorithms That Changed the Future: The Ingenious Ideas That Drive Today's Computers*. New Jersey: Princeton University Press.

Chapter 6

1) Humans and Anatomy Lab. See their research agenda at https://hal.pratt.duke.edu/research-0

2) Ibid.

3) Osbourne, Michael A., and Carl Frey (2013). *The Future of Employment: How Susceptible are Jobs to Computerisation*. Oxford University. September 2013. Available from: http://

www.oxfordmartin.ox.ac.uk/downloads/academic/The_Future_of_Employment.pdf [accessed 19.04.17]

4) Ibid; Osbourne, Michael A., and Carl Frey. Technology at Work: The Future of Innovation and Employment. Citi GPS February 2015. http://www.oxfordmartin.ox.ac.uk/downloads/reports/Citi_GPS_Technology_Work_2.pdf [accessed 19.04.17]

5) Artificial intelligence: The impact on jobs; Automation and anxiety. *The Economist.* June 2016. Available from: http://www.economist.com/news/special-report/21700758-will-smarter-machines-cause-mass-unemployment-automation-and-anxiety [accessed 19.04.17]

6) Graeber, David. On the Phenomenon of Bullshit Jobs. *Strike Magazine.* August 2013. Available from: http://strikemag.org/bullshit-jobs/ [accessed 19.04.17]

7) Weir, Kirsten. Never a dull moment. *American Psychological Association.* July/August 2013, Vol 44, No. 7.

8) Benjamin, Andrew (2013). *Working with Walter Benjamin: Recovering a Political Philosophy.* Edinburgh: Edinburgh University Press. p. 212.

9) Koerth-Baker, Maggie. Why Boredom Is Anything but Boring. *Scientific American.* January 2016. Available from: https://www.scientificamerican.com/article/why-boredom-is-anything-but-boring/ [accessed 19.04.17]

10) Goodstein, Elizabeth (2004). *Experience Without Qualities: Boredom and Modernity.* Stanford, Ca.: Stanford University Press.

11) Prinzel, Lawrence J., III, DeVries, Holly, Freeman, Fred G., Mikulka, Peter (2001). Examination of Automation-Induced Complacency and Individual Difference Variates. NASA Langley Research Center; Hampton, VA United States. December 2001.

12) A Review of Flightcrew-Involved Major Accidents of U.S. Air Carriers, 1978 Through 1990. National Transportation

Safety Board. January 1994.
Available from: http://libraryonline.erau.edu/online-full-text/ntsb/safety-studies/SS94-01.pdf [accessed 19.04.17]

13) Cleary, M. J. Sayers, V Lopez, C. Hungerford. Boredom in the Workplace: Reasons, Impact, and Solutions. *Issues in Mental Health Nursing*. 2016;37(2):83-9.

Chapter 7

1) Anderson, Chris (2008). The End of Theory: The Data Deluge Makes the Scientific Method Obsolete. *Wired*. June. Available from: https://www.wired.com/2008/06/pb-theory/ [accessed 19.04.17]

2) Dwoskin, Elizabeth (2017). The next hot job in Silicon Valley is for poets. *The Washington Post*. April. Available from: https://www.washingtonpost.com/news/the-switch/wp/2016/04/07/why-poets-are-flocking-to-silicon-valley/?utm_term=.3ed152b7b4ce [accessed 19.04.17]

3) Ibid.

4) Lohr, Steve (2013). Algorithms Get a Human Hand in Steering Web. *The New York Times*. March 2013. Available from: http://www.nytimes.com/2013/03/11/technology/computer-algorithms-rely-increasingly-on-human-helpers.html [accessed 19.04.17]

5) Lieberman, Philip (2013). *The Unpredictable Species: What Makes Humans Unique.* Princeton, N.J.: Princeton University Press.

6) González, Marta C., César A. Hidalgo and Albert-László Barabási (2008). Understanding individual human mobility patterns. Nature 453, 779-782 (5 June).

7) Ibid.

8) Alex Pentland and Andrew Liu (1999). Modeling and prediction of human behavior. *Neural Computation*. 11, 1 (January), 229-242.

9) Qin S-M, Verkasalo H, Mohtaschemi M, Hartonen T, Alava

M (2012). Patterns, Entropy, and Predictability of Human Mobility and Life. PLoS ONE 7(12): e51353.

10) Bagrow JP, Lin Y-R (2012). Mesoscopic Structure and Social Aspects of Human Mobility. PLoS ONE 7(5): e37676. doi:10.1371/journal.pone.0037676

11) Nathan Eaglea, Alex (Sandy) Pentland, David Lazer (2009). Inferring friendship network structure by using mobile phone data. *Proceedings of the National Academy of Sciences of the United States of America.* Vol 106, no. 36.

12) Ibid.

13) Won, A.S., Bailenson, J.N., Stathatos, S.C. et al (2014). Automatically Detected Nonverbal Behavior Predicts Creativity in Collaborating Dyads. *Journal of Nonverbal Behavior.* Nonverbal Behav (2014) 38: 389. doi:10.1007/s10919-014-0186-0

14) Ibid.

15) Harari, Yuval (2017). *Homo Deus: A Brief History of Tomorrow.* New York: Harper Collins.

16) Harari, Yuval. Yuval Noah Harari on the Rise of Homo Deus. Interview by Intelligence Squared. September 2016. Available from: http://www.intelligencesquared.com/events/yuval-noah-harari-on-the-rise-of-homo-deus/ [accessed 19.04.17]

17) Jones, Bryan D (2003). "Bounded Rationality and Political Science: Lessons from Public Administration and Public Policy." Journal of Public Administration Research and Theory: J-PART 13, no. 4 (2003): 395-412. http://www.jstor.org/stable/3525655 [accessed 19.04.17]

18) Jones, Cliff, Lloyd, John L., eds (2011). *Dependable and Historic Computing Essays Dedicated to Brian Randell on the Occasion of his 75th Birthday.* New York: Springer-Verlag Berlin Heidelberg.

19) Jhingran, Anant (2016). Obsessing over AI is the wrong way to think about the future. *Wired.* January 2016. Available

at: https://www.wired.com/2016/01/forget-ai-the-human-friendly-future-of-computing-is-already-here/ [accessed 19.04.17]

20) Arbesman, Samuel (2014). It's Complicated. *Aeon. com*. January. Available from: https://aeon.co/essays/is-technology-making-the-world-indecipherable [accessed 19.04.17]

21) Hillis, Danny (2016). The Enlightenment is Dead, Long Live the Entanglement. *Journal of Design and Science*. MIT press. Available from: http://jods.mitpress.mit.edu/pub/enlightenment-to-entanglement. [accessed 19.04.17]. See also Arbesman, Samuel (2014).

22) Neff, Gina, Tim Jordan, and Joshua McVeigh-Schulz (2012). Affordances, technical agency, and the politics of technologies of cultural production. *Culture Digitally* [see also their blog]. January 2012. Available from: http://culturedigitally.org/2012/01/affordances-technical-agency-and-the-politics-of-technologies-of-cultural-production-2/ [accessed 19.04.17]

23) Wissner-Gross, Alex (2014). See his Ted Talk, recorded on 2014. Also see his website with various publications at http://www.alexwg.org/

24) Wissner-Gross, Alex (2015). Engines of Freedom. *Edge.com*. Available at: https://www.edge.org/response-detail/26181 [accessed 19.04.17]

Chapter 8

1) Kelly, Kevin (2915). Why I Don't Worry About a Super AI. The Technium. April 2015. Available from: http://kk.org/thetechnium/why-i-dont-worry-about-a-super-ai/ [accessed 19.04.17]

2) Big Data: The New Natural Resource. IBM. See IBM's website, especially http://www.ibmbigdatahub.com/infographic/big-data-new-natural-resource [accessed 19.04.17]

3)   Lechner, Rich. A Closer Look at IBM's "Smarter Planet" Campaign. Interview conducted by Joel Makower, December 2008, and found at: https://www.greenbiz.com/news/2008/12/30/closer-look-ibms-smarter-planet-campaign [accessed 19.04.17]

4)   Hewlett Packard. Quote pulled from their corporate website, post titled CeNSE. Available from: http://www8.hp.com/us/en/hp information/environment/cense.html#.WOEMLBLys_U [accessed 19.04.17]

5)   Duane, Daniel. The Unnatural Kingdom: If technology helps us save the wilderness, will the wilderness still be wild? *The New York Times*. March 2016. Available from: https://www.nytimes.com/2016/03/13/sunday-review/the-unnatural-kingdom.html [accessed 19.04.17]

6)   Macmanus, Richard. Chinese Premier Talks Up Internet of Things. *Readwrite.com*. January 2010. Available from: http://readwrite.com/2010/01/19/chinese_premier_internet_of_things/ [accessed 19.04.17]

7)   Navlakha S, Bar-Joseph Z. Algorithms in nature: the convergence of systems biology and computational thinking. *Molecular Systems Biology*. 2011;7:546. doi:10.1038/msb.2011.78.

8)   Kurtzweil, Ray (2006). Reinventing humanity: The future of human-machine intelligence. *The Futurist*. March-April 2006. Can be found online at: http://www.kurzweilai.net/reinventing-humanity-the-future-of-human-machine-intelligence [accessed 19.04.17]

9)   The urgency of questions can be illustrated rather easily. A man's house is on fire. Inside are his children, wife, two pets and all his belongings; however, said man is busy pruning the tree out front, since, after all, if the house burns the tree will remain, and he might as well have a nice looking tree left than no tree at all. He is worried about his property once the house is gone. The man exhibits decent logic here,

but it is without value. It is not that there are right or wrong questions per se, but timely questions. True, this man is thinking ahead, and he has a point, but his reasoning is anchored in what many would understand to be misguided concern—the tree. He should be thinking about saving his children, wife and pets. He is asking a good question, but at the wrong time. He just can't prioritize, and his reasoning, which is an activity by itself, cannot provide a priority because it has no end and no beginning; science needs ideology—without it, science, like all action, is blind. It does not matter if a boxer has a punch that is twice as strong as his opponent's—if his opponent knows how to hit at just the right time, only then will the punch matter. If we were a species with wings, our scientific inquiries would revolve around an entirely different set of suppositions, and rather than put so much science into flying and space exploration, we might be feverishly be trying to figure out how to live on the flat ground. As the human mind evolves with its surroundings, much as it did when we were hunter-gatherers, then as pastoralists, then as seafaring people, and so on, we are really asking what we are doing and why we are doing it. The way we study intelligence will never be independent from the modalities through which intelligence is tested. In our biotechnological age, we find the most promising analysis of intelligence to be from neuroscientists, but two thousand years ago the best guess as to how intelligence works might come from the Delphic oracle or an unemployed philosopher.

Chapter 9

1) Griffin, Andrew. Tech billionaires convinced we live in the Matrix are secretly funding scientists to help break us out of it. *Independent*. October 2016. Available from: http://www.independent.co.uk/life-style/gadgets-and-tech/news/

computer-simulation-world-matrix-scientists-elon-musk-
artificial-intelligence-ai-a7347526.html [accessed 19.04.17]

2) Wilson, Edward (1984). *Biophilia*. Cambridge, MA: Harvard University Press.

3) Wilson, Edward (2010). *The Diversity of Life*. Cambridge, MA: Belknap Press. p. 350.

4) Thomas, Sue (2013). *Technobiophilia Nature and Cyberspace*. New York: Bloomsbury.

5) See the discussion in *The Biophilia Hypothesis*, ed. by Stephen R. Kellert. Washington D.C.: Island Press; 1993. p. 84.

6) Ibid.

7) Harari, Yuval (2016). Homo sapiens is an obsolete algorithm: Yuval Noah Harari on how data could eat the world. *Wired*. September 2016. Available from: http://www.wired.co.uk/ article/yuval-noah-harari-dataism [accessed 19.04.17]

8) Marenko, Betti (2014). Neo-Animism and Design: A New Paradigm in Object Theory. *Design and Culture*. Volume 6, issue 2. Available at: https://architecture.mit. edu/sites/architecture.mit.edu/files/attachments/lecture/ MARENKO_Neo-Animism%20and%20Design.pdf [accessed 19.04.17]

9) Davis, Erik (2015). *TechGnosis: Myth, Magic, and Mysticism in the Age of Information*. Berkeley, Ca.: North Atlantic Books.

Chapter 10

1) Mori, M (2012). The Uncanny Valley. Translated by MacDorman, K. F.; Kageki, Norri. IEEE Robotics and Automation. 19 (2): 98–100.

2) Kelly, Kevin (2015). Why I Don't Worry About a Super AI. *The Technium*. April 2015. Available from: http:// kk.org/thetechnium/why-i-dont-worry-about-a-super-ai/ [accessed 19.04.17]

In the Dust of This Planet
Horror of Philosophy vol. 1
Eugene Thacker
In the first of a series of three books on the Horror of Philosophy, In the Dust of This Planet offers the genre of horror as a way of thinking about the unthinkable.
Paperback: 978-1-84694-676-9 ebook: 978-1-78099-010-1

Capitalist Realism
Is there no alternative?
Mark Fisher
An analysis of the ways in which capitalism has presented itself as the only realistic political-economic system.
Paperback: 978-1-84694-317-1 ebook: 978-1-78099-734-6

Romeo and Juliet in Palestine
Teaching Under Occupation
Tom Sperlinger
Life in the West Bank, the nature of pedagogy and the role of a
university under occupation.
Paperback: 978-1-78279-637-4 ebook: 978-1-78279-636-7

Sweetening the Pill
or How we Got Hooked on Hormonal Birth Control
Holly Grigg-Spall
Has contraception liberated or oppressed women? *Sweetening
the Pill* breaks the silence on the dark side of hormonal
contraception.
Paperback: 978-1-78099-607-3 ebook: 978-1-78099-608-0

Why Are We The Good Guys?
Reclaiming your Mind from the Delusions of Propaganda
David Cromwell
A provocative challenge to the standard ideology that Western
power is a benevolent force in the world.
Paperback: 978-1-78099-365-2 ebook: 978-1-78099-366-9

Readers of ebooks can buy or view any of these bestsellers by
clicking on the live link in the title. Most titles are published
in paperback and as an ebook. Paperbacks are available in
traditional bookshops. Both print and ebook formats are
available online.

Find more titles and sign up to our readers' newsletter at http://
www.johnhuntpublishing.com/culture-and-politics
Follow us on Facebook at
https://www.facebook.com/ZeroBooks
and Twitter at https://twitter.com/Zer0Books